JK   Rae, Nicol C.   28,343
2316
.R34 Southern Democrats
1994

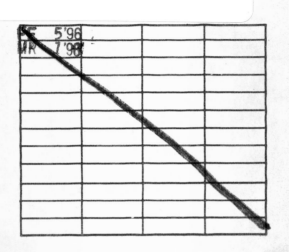

Southern Democrats

Southern Democrats

NICOL C. RAE

New York   Oxford
Oxford University Press
1994

Oxford University Press

Oxford   New York   Toronto
Delhi   Bombay   Calcutta   Madras   Karachi
Kuala Lumpur   Singapore   Hong Kong   Tokyo
Nairobi   Dar es Salaam   Cape Town
Melbourne   Auckland   Madrid

and associated companies in
Berlin   Ibadan

Copyright © 1994 by Oxford University Press, Inc.

Library of Congress Cataloging-in-Publication Data
Rae, Nicol C.
Southern Democrats / Nicol C. Rae.
p.   cm.
Includes bibliographical references and index.
ISBN 0-19-508708-9.
ISBN 0-19-508709-7 (pbk.)
1. Democratic Party (U.S.)
2. Southern States—Politics and government—1951-
3. United States—Politics and government—1945-1989.
4. United States—Politics and government—1989-
I. Title.
JK2316.R34   1994   324.275′06—dc20   93-32876

2 4 6 8 9 7 5 3 1

Printed in the United States of America
on acid-free paper

Dedicated to the Memory of
My Father,
John Nicol Rae
1936–1992

PREFACE

These pages emerged from a long-standing personal interest in the history and culture of the South, and my academic interest in the evolution of American political parties. As it transpired, the southern perspective proved to be a fascinating vantage point from which to view the transformation of American national politics, and more particularly Democratic party politics, over the past quarter-century or so. The southern Democrats—a regional intraparty faction—achieved almost a dominant position in national politics in the years immediately following the Second World War, through their grip on the United States Congress, a hold that no president—Democrat or Republican—proved capable of breaking.

Yet even at this time, forces were working below the surface of American politics that undermined the southerners' power: economic change, demographic change, the communications revolution, and finally the emerging challenge to the southern social and political system by the black southerners who had been deliberately excluded from it. The *raison d'être* of southern Democrats in national politics prior to the civil rights revolution was to preserve the southern caste system of racial segregation and disfranchisement. When this was dismantled by the federal courts, the civil rights movement, and the Johnson administration, it appeared that the day of southern Democrats in national politics was over when the Republican party of Goldwater, Nixon, and Reagan took over the old white segregationist constituency and the Democrats appeared likely to be left (ironically) with the votes of the newly enfranchised black minority but little else.

In presidential elections this is more or less what occurred. Virtually nonexistent in the South before the civil rights revolution, the Republicans became dominant in presidential elections in the South after 1968. But in congressional, state, and local races, the Democratic party not only did not slide into a minority position, but it also maintained a considerable superiority throughout the region. This durability and persistence of Democratic power below the presidential level was the main factor in aborting Republican hopes since 1968 for the overall realignment of American politics.

This book seeks to explain this conundrum and to analyze just how and to what extent southern Democrats have changed since the civil rights revolution. Despite all that has passed—the nationalization of the South, economically, culturally, and politically—and the reduced distinctiveness of southern Democrats from national Democrats, the southerners nevertheless have retained a pivotal position in national politics. If the Democrats cannot at least split the South with the Republicans in presidential elections, they are likely to lose, and in Congress it was the votes of southern Democrats that invariably decided the outcome in most of the major congressional battles of the Reagan–Bush era. Therefore to understand what has been happening to the southern Democrats and what is likely to happen to them during the 1990s is also to understand what has been happening in American politics more generally in these times.

Much of the inspiration for this book came from Prof. Byron E. Shafer of Nuffield College, Oxford, a constant source of insight, wisdom, and motivation since I first came to know him as his graduate student some years ago. Alex Lamis gave me the benefit of his advice in approching southern politics and generously provided useful materials for my research. Ron Peters, Patricia Hurley, Howard Reiter, and Hall Bass rendered penetrating and encouraging comments on various portions of the manuscript. Jon Hale shared his insights from his research on the Democratic Leadership Council, and was an excellent sounding board for ideas on Democratic intraparty politics. David Mayhew, Norman Ornstein, James Reichley, David Butler, Richard Neustadt, Tom Langston, John and Kate Galbraith, and Andrew Adonis should also be thanked for consistently providing encouragement, sound advice, and amicability. Finally, I am especially indebted to Michael E. Lind, a constant source of intellectual inspiration over the years.

In Miami, my colleagues in the political science department at Florida International University provided a stimulating environment where the (mercifully) animated discussion of contemporary political events has only rarely been subordinated to professional and bureaucratic considerations. I am particularly grateful to Dario Moreno, John Stack, Mary

Volcansek, and Joel Gottlieb for their hospitality and reassurance during the writing of this book. Darden Ashbury Pyron gave me the benefit of his wit and erudition with regard to southern history and culture, and Peter Craumer, Camilla Guido, and Dorinda Mosby helped me get through various logistical and technological crises that were largely of my own making. One more debt of gratitude is owed to Miguell Del Campillo for showing me a life and world beyond academe that was more than worthy of my attention.

This is also the proper place to thank all those persons who allowed themselves to be interviewed and whose observations on the state of American national politics in the early 1990s were essential to the writing of this book.

Finally, I am indebted to my late father and to my mother in more ways than mere language can express, but particularly for teaching me that "everyday life" can be as heroic, if not more so, as the "grand arena" of politics and diplomacy.

Miami Beach   N.C.R.
August 1993

CONTENTS

LIST OF TABLES

Southern Democrats

Introduction

This book is about the South, and particularly southern Democrats, in national party politics since the culmination of the civil rights revolution in 1965 (the year that the Voting Rights Act was passed).

The South has been the major aberration in the pattern of America's economic and social development since the Civil War, and as a consequence, southern party politics also has deviated from the national norm. The clearest illustration of this deviation was the establishment of the one-party Democratic "Solid South" at both the state and national levels in order to maintain the South's hierarchical social system based on racial segregation and black disfranchisement. Since the passage of the Voting Rights Act in 1965, the old barriers of segregation and disfranchisement have disappeared, and rapid economic development has brought an end to the old southern economic and social system. One outcome of these changes has been the emergence of a competitive two-party system in the South.

Yet despite all this, the southern Democrats remain distinctive, even though the civil rights issue—which was thought to constitute the sole rationale for their deviation—no longer appears to arouse significant divisions in the national party. Since the late 1960s the persisting Democratic advantage at all levels of southern electoral politics below the presidential level has also successfully aborted the Republicans' efforts to realign the national party system. Moreover, the survival of the southern Democrats and the degree of cohesion that they have maintained, particularly in Congress, give us an excellent perspective from which to examine the evolution of the new polarized factional politics in American parties.

The southern Democrats are the sole example of a nonideological faction based on local and regional concerns surviving in a modern American political environment that is largely antipathetic to this style of politics.

Whereas other recent accounts of southern politics have tended to concentrate on the state or local level, this book focuses on the national context and the place of the southern Democrats in that context. It is apparent that Democratic control of the major national political institutions — Congress and the presidency — is contingent on the continued allegiance of the southern Democrats to the national party. The nomination and election of two southern Democrats as president and vice-president of the United States in 1992 highlight the continuing significance of the South to the party's national success.

My research is based on interviews with significant actors in the process, generally available data, and memoirs and secondary accounts of past campaigns. The approach of the book is narrative and chronological rather than quantitative, though numerical data are utilized where appropriate.

The book is divided into seven chapters:

Chapter 1 introduces the subject and places the southern Democrats in the broader context of American party factionalism, an understudied topic to which I have already devoted considerable attention in my work on liberal Republicans. I describe the southern Democrats as an aberration in the development of a more ideological pattern of intraparty factionalism in the United States and discuss how and why they have been able to survive and even prosper in this context.

Chapter 2 summarizes the history of the old southern Democracy and the racial politics that sustained it from the Civil War to 1965. The strategies used by the southerners to maintain their grip on the national party, even after the New Deal, are also discussed, together with the reasons for the final collapse of the old one-party Democratic South in the mid-1960s.

Chapter 3 looks at the transformation of the Democratic presidential party after 1968, from the perspective of the southern Democrats, and shows how the new primary-based presidential nominating process initially eroded their influence. I explore the exceptional case of Jimmy Carter in 1976, along with more recent efforts to reassert the South's position, such as that in "Super Tuesday" in 1988.

Chapter 4 examines the Democratic party in Congress and the erosion of southern Democratic influence in both the House and the Senate.

This chapter is based largely on interviews with southern Democratic congressmen, senators, and staffers in the summer of 1990.

Chapter 5 looks at the Democratic Leadership Council (DLC) and the (predominant) role of the southern Democrats in its formation. I look at the DLC's aims and strategy as well as its organizational structure and sources of support. Again, this section relies heavily on interviews with senior DLC operatives in the summer of 1990.

Chapter 6 deals with the success of southern Democrat and DLC leader Bill Clinton in winning the 1992 Democratic presidential nomination, contrary to the pattern of recent presidential nominating contests, which have not favored "moderate" candidates. The chapter examines how Clinton was able to win the nomination from his more liberal rivals and the extent to which he has been able to transform the direction of the party. I also describe the reasons for the success of the Clinton–Gore ticket in the general election.

Chapter 7 concludes with an overall assessment of the present and future state of the southern wing of the Democratic party and some final reflections on the nature of contemporary American party politics and party factionalism.

1

The South and American Party Factionalism

Intraparty factionalism is one of the most understudied areas of American political science. Although we have strong evidence of increasingly polarized, coherent, and durable party factions in both the national Republican and Democratic parties, an overall analysis of American party factionalism is still lacking. The remainder of this chapter is an attempt to arrive at such an analysis by looking at the nature of factionalism in the Republican and Democratic parties today.

Factions and Factionalism

Before embarking on this task, however, it is necessary to consider what precisely is meant by the term *faction*. In classical republican usage, *faction* was almost always used pejoratively to refer to self-interested groups of persons who sought to manipulate the polity for their own ends rather than the common good. As such, factions were generally condemned as a danger to the health of the republic as a whole.[1] This usage of the term was still current when critics of the British government in the early eighteenth century led by Lord Bolingbroke condemned the patronage-oriented "factional" politics of Robert Walpole's regime.[2] Indeed, in his remarks on the evils of factionalism, Bolingbroke made no distinction between parties and factions, seeing the latter as simply the most degenerate case of the former:

For faction is to party what the superlative is to the positive: party is a political evil, and faction is the worst of all parties . . . factions are in them [societies], what nations are in the world; they invade and rob one another: and while each pursues a separate interest, the common interest is sacrificed by them all: that of mankind in one case, that of some particular community in the other.[3]

The framers of the American Constitution continued to see factions and parties as indistinguishable and to deplore the "mischiefs of faction" in regard to the well-being of the commonwealth. Nevertheless, unlike Bolingbroke and the Tory critics of the English constitution, they did not believe that factions could be extinguished from government without violating the integrity of free government itself. The deleterious effects of factions, however, might be mitigated by establishing a governmental system based on separated institutions sharing essential powers, so that no single faction would be able to predominate.[4]

By the early nineteenth century the terms *party* and *faction* were no longer used synonymously. With the advent of political parties as crucial intermediaries between mass electorates and government, they could no longer be seen as merely self-interested cliques of office seekers but, rather, as associations of individuals that aimed at controlling government in the name of some common idea or opinion. According to Sartori, the transition is evident in the more positive references to parties in the works of Hume and Burke in the late eighteenth century.[5] With this change in terminology, *factions* came to refer to organized groups within political parties, although the term did retain its pejorative connotations.

This definition, however, was too broad to satisfy Sartori, who pointed out that it fails to distinguish among different kinds of party subunits (e.g., local and/or regional branches or ideological groups).[6] Sartori also follows Richard Rose in differentiating between factions, which are "self-consciously organized . . . with a measure of discipline and cohesion," and tendencies, which are "a stable set of attitudes rather than a stable group of politicians" lacking both the organization and continuity of personnel characteristic of factions.[7] He thus prefers the term *fractions* as a generic term to refer to party subunits, factions, and tendencies.[8] Although I appreciate the differences in degree among factions in different parties and party systems, I do not see tendencies as being intrinsically different from factions, but merely as a more weakly developed

species of the latter. Thus in this discussion of American party factionalism, I shall use the term *faction*.

In general, party factions seem to fall into two main categories:
clientelistic and ideological.[9]

Clientelistic factionalism appears to be characteristic of center-right
political parties, whose factional divisions are oriented around the
factions' ability to deliver patronage or favors for their specific constituency or "clientele." These factions are powerful and disciplined,
with a high degree of continuity. They frequently originate from the
personal following of a particularly charismatic or important party
leader, but the leadership of these factions is transferred from one
generation of leaders to another, sometimes within the family of the
founder. If the faction is of long standing, it is generally officially or
unofficially "institutionalized" in the organizational structure of the
"host" party. Clientelistic factionalism has been most characteristic of
party systems in underdeveloped societies or "dominant-party" systems
in which a single center-right party, such as the Italian Christian Democrats, the Japanese Liberal Democrats, or the Radicals in the French
Third Republic, has an almost permanent grip on governmental power.[10]

Ideological factions are more characteristic of Socialist parties or
parties on the left of the political spectrum, in which ideology plays a
much wider role than in center-right parties. Here the factional differences are grounded not in patronage or charisma but in fundamental
differences of opinion within the party's ranks over the policies the party
should adopt and how much the party's commitment to the socialist ideal
should be emphasized. The British Labour party, the French Socialist
party, and the German Social Democratic party all display features of this
type of factionalism. Given that the nature of the factional demands in
this case are ideological rather than material, ideological factional divisions are much more difficult for parties to accommodate, and so left-
wing parties in the contemporary political world generally have suffered
much more from prolonged and irreconcilable intraparty divisions than
have their more conservative opponents. By contrast, clientelistic factional divisions are almost a positive asset for the dominant parties, since
the variety of factions enables the governing party to accommodate a
much greater variety of demands. In dominant-party systems, the diversity of factionalism within the dominant party enhances the party's
representativeness and thus partly compensates for the lack of rotation in
government between parties.

American Party Factionalism Before the New Deal

American political parties are unique. Whereas the tendency of political parties in other liberal democratic systems during the first half of the twentieth century was toward more ideological, class-based, centralized, and disciplined party organizations, the American parties remained in a largely premodern state: nonideological, undisciplined, and decentralized, manifesting themselves as national entities only during presidential elections.

Lacking strong, centralized parties, the United States also lacked strong and disciplined party factions (with the possible exception of the southern Democrats in their heyday). Roback and James argued that factionalism had been prevalent in the machine politics of the nineteenth century, with its concomitant intraparty struggles over spoils and patronage, but had withered as the parties themselves had withered during the twentieth century. They argued that American party factions in the modern era were really "tendencies" (in Rose's terminology), too ephemeral and unstructured to merit classification as factions:

> The most common tendencies are candidate-centered party organizations that wage political battle and then usually disintegrate. Often such tendencies have ideological roots because they represent various ideological wings of the party. Tendencies often are linked to political subculture and vary by region, state, or locality. The linking of a candidate's campaign organization with an ideological mood constitutes the most common form of tendency in current American politics.[11]

This view of American party factions as tendencies is largely shared by Reiter, but he sees more continuity in the personnel of American party factions than do Roback and James, and he prefers to use a third term, *cluster*, to describe the nature of American party factions:

> Nevertheless we would maintain that we are describing stable groups of politicians, for certain names come readily to mind as obvious candidates for each "tendency" and there is at least a bit of organization. At the risk of needlessly proliferating terminology, we prefer to use a third term, "cluster," which expresses the identifiability of individuals without implying organizational coherence.[12]

As I stated earlier, my own preference is to retain the original term, *faction*, largely for reasons of clarity but also because American party factions have become less amorphous and more durable in the past decade or so.

At least since the Civil War there has been a recurring pattern of intraparty cleavages in the national Democratic and Republican parties. Presidential nominating contests between the Civil War and New Deal revealed that both national parties, in Congress and at national party conventions, generally divided on a regional basis: Northeast/West/ South in the case of the Democrats, and East/West in the case of the Republicans.[13] Analyses of congressional voting over the same period by Bensel and Sinclair reveal a similar pattern.[14]

The factors underlying this factional division on regional lines were both economic and cultural. The party system itself primarily reflected the national cleavage created by the Civil War. In the South the Democratic party was completely dominant, white southern elites having established a virtual one-party system that largely excluded blacks and low-status whites from political participation after the end of Reconstruction. Outside the South the party was generally in a minority except in some border and western states. The Republicans were the dominant national party because of their strength in the industrial North and the agrarian West.

In many respects the Civil War's cleavage reflected deeper cultural divisions. In those parts of the country that had been settled by New England Yankees or northern Europeans, Protestant immigrants tended to be Republican, whereas those areas that had been settled by south- erners or mountain whites, or by non-Protestant immigrant groups, tended to be Democratic.[15]

The economic cleavage between the northeastern metropolitan areas and the southern and western agrarian hinterland – between the industrial core and the periphery – further complicated the picture.[16] The presence of the Roman Catholic immigrant groups in the cities gave the Democrats the chance to win support in the core, but to do so they had to downplay the "southern" domination of the party and to try to appeal to the northern working class in class terms. With the exception of Grover Cleveland and Woodrow Wilson, they failed.

The Republicans also feared that the western section of the party would be seduced by the radical politics of the populists and progressives, because of the endemic crisis in American agriculture after the Civil War. This fundamental incompatibility in economic interests between the eastern and western sections of the party was a serious problem for the Republicans, though they were generally able to get around it by refocusing the debate on cultural issues. They were also assisted by the absorption into the Democratic party of much of the populist and radical sentiment of William Jennings Bryan, whose revivalist political style discomfited most midwestern and western Republicans.[17]

It was these intraparty economic and cultural tensions that gave rise to a recurring pattern of sectional division in both parties. Although the eastern and western wings of the Republican party contended for influence during this period, the eastern financial and industrial wing generally triumphed.[18] The Republicans' western radical wing had its greatest strength in Congress, and particularly in the U.S. Senate, where the overrepresentation of the smaller western states increased their numbers.

In regard to the Democratic party, the South held a commanding position in the congressional party, since it was the strongest Democratic region, but with memories of the Civil War still fresh, the election of a southerner to the presidency (unless like Woodrow Wilson, he had been de-southernized) was inconceivable. In these circumstances the Democrats had to look outside the South for a nominee, but that nominee would nonetheless have to be acceptable to the South insofar as he pledged not to interfere with the southern social and economic caste system. The southern veto on nominees was institutionalized by the two-thirds rule at Democratic National Conventions. Finding a nominee with electoral appeal in the North and West who was also acceptable to the South proved to be an extremely difficult task for the Democrats before the New Deal, particularly in the disastrous elections of 1924 (when the party took 103 ballots to find a nominee) and 1928 (when they nominated the New York Irish Catholic Al Smith and lost half of the South and all of the West).[19]

These pre–New Deal American party factions were not tightly organized or disciplined, but unlike Rose's tendencies, the recurring sectional divisions within the parties did have continuity beyond one or two elections. These intraparty cleavages also appear to have been peculiarly American, since they fit neither the clientelistic nor the ideological categories described earlier. The factional divisions appeared with particular virulence when an issue arousing sectional cleavages within the parties came to the fore, such as the tariff in the case of the Republicans or "free silver" in the case of the Democrats. Yet there was no coherent program that each party faction represented, nor — at the national level — were durable factions formed around particular party leaders or families. Rather, traditional American party factions reflected persisting economic and cultural differences among geographical sections of the country.[20]

The selection of the party's presidential nominee was in the hands of the patronage-oriented party leaders or bosses who controlled the national convention delegations of the major parties: In short, the party's factional conflict was managed by party elites who had to sell a compromise candidate or platform to their followers. These elites had a vested interest in the party's national electoral success, since it would guarantee them access to federal patronage. It is in this light that we should understand the often prolonged balloting, "favorite son" candidacies, demonstrations, and "stampedes" that characterized the national party conventions of the pre–New Deal period. The inevitable "balancing" of the party's presidential "ticket," with the vice-presidential nomination going to someone from the losing faction, was an additional feature of the "sectional" American party factionalism of this period.

When the conflict between the sections involved economic questions, the task of the state and regional party leaders was fairly straightforward, since on these matters they could split the difference between factional demands and find compromise candidates or party platforms to assuage their followers. When the sectional cleavages arose around ethical and/or cultural questions, however, as in the Democratic presidential marathon of 1924, the divisions were all but impossible to heal.

In the pre–New Deal period, then, we have evidence of persistent factionalism in both the major American parties, grounded in recurring sectional differences. These factions were loose and relatively undisciplined and had little organizational continuity, but they were grounded in deep-rooted and durable sectional loyalties at the mass level. The factions were not "ideological" in the sense of promoting a doctrine or worldview, nor at the national level could they accurately be described as "clientelistic." Instead, factional competition was managed by regional and state party leaders who tried to negotiate compromises at national party conventions that would keep the national parties together. When the factional conflict proved too bitter for the bosses to contain (as in the Republican party in 1912 or among the Democrats during the 1920s), the divided party went down to ignominious defeat.

The New Deal, a Period of Transition

Since the New Deal, American party factionalism has been increasingly based on ideological rather than sectional divisions within the parties.

Sectional factors have by no means disappeared, of course, but the emphasis of the factional conflict has been more on ideology and policy rather than sectional issues. The New Deal era (1932–68) temporarily replaced a party system based on culture with a system based largely on socioeconomic class—outside the South. Although the New Deal Democrats created a system of government intervention and welfare provision to benefit labor and the northern working class, they left the southern racial system untouched. Southerners in Congress reciprocated by allowing the passage of significant class-oriented legislation that did not impinge on their fief.[21] The evolution of the committee seniority system in the Democratic-dominated Congresses institutionalized the system and maintained the key position of the southern wing in the party. Indeed, the southern Democrats—now a minority faction in the congressional party—manipulated the system so as to achieve a dominant position in the Congress during the New Deal era.[22]

The southerners were thus the major element in the Democratic party's "conservative" faction during this period. The party's northern New Dealers, based in organized labor and the major state and local party organizations of the northern states, were generally regarded as the "liberal" faction. The liberals were the dominant faction in presidential politics, and their domination was assisted by the abolition of the two-thirds rule at President Franklin Delano Roosevelt's behest in 1936. However, they still needed southern Democratic support in order to win the presidency and to pass their programs in Congress. They thus acceded to the southerners' domination of the Congress and their virtual veto on civil rights legislation.

The Republican party, now in the minority position nationally, also divided into two factions, and here the factional competition in the New Deal years was much more ferocious than among the Democrats. During the New Deal the Republicans' radical, western, agrarian wing was largely absorbed into the Democratic party, but this did not end sectional conflict within the GOP. The party's "stalwarts," the regular Republicans in the small towns and medium-sized cities of the Northeast and Midwest, bitterly resented the New Deal and its interventionist policies at home and abroad. Urban, corporate, Republicans on the East Coast, on the other hand, found accommodation with the New Deal to be much easier and in accordance with their own economic interests.

By 1940 the "Wall Street" or "eastern establishment" wing of the party had made its peace with the New Deal, when they secured the presidential

nomination for the unknown maverick Wendell Willkie in preference to
the stalwart favorite Senator Robert Taft. During the New Deal era the
two wings of the party warred over the party's presidential nomination,
with the urban, Wall Street wing of the party in the ascendancy. In
Congress, however, the stalwarts dominated, reflecting the party's congressional base in rural and suburban, northern, white Protestant America. During the New Deal period, the degree of factional consciousness
between the two factions was extremely high by American standards, and
the factions became increasingly well organized and established. Interestingly, the Republican party's factions increasingly lost their geographical focus (though not entirely) and competed with one another at all
levels of the party nationwide.[23]

As the New Deal party system evolved, the factional pattern in the
parties began to change. In the Democratic case, the "liberal" northern
faction began to show signs of fragmentation as early as the late 1940s,
when it became apparent that many of its adherents wanted to move the
party more rapidly toward an explicitly pro–civil rights position. The
inclusion of a civil rights plank in the 1948 party platform precipitated the
bolt of Strom Thurmond and the southern hard-liners in that year, though
most of the South remained loyal to Harry Truman in November.

During the 1950s, tension in the liberal wing became more apparent.
Younger active Democrats in the North and West became increasingly
impatient with the cautious, pluralistic approach to politics of the New
Deal liberals who dominated the party, and they sought to emphasize a new
politics of "participation" and "reform." Many of the reformers were
attracted to party politics by the party's idiosyncratic presidential nominee
of the 1950s, Adlai Stevenson, and in 1960 they were enthusiastic supporters of John Kennedy against more traditional New Dealers such as Lyndon
Johnson and Hubert Humphrey. These people were not traditional northern blue-collar or "ethnic" Democrats but, rather, enjoyed a middle-class
life-style and had professional occupations. Largely employed in the
talking, writing, and thinking professions, their focus was more on political style, single issues, and ideology, rather than the basic "haves versus
have-nots" politics of the New Deal. By the early 1960s, they were already
approaching an ascendant position in the northern section of the party.[24]

The New Deal era thus was a period of transition from a sectional
pattern of American party factionalism toward a more ideological type of
conflict, although sectional influences still operated in both parties. In
the Republican party, the old East/West cleavage was becoming subordi-

nated to a conflict between "conservatives" and "moderates," preoccupied more with matters of style and ideology than sectional concerns. Similarly, in the Democratic party, although a strong regional faction persisted in the South, among the northern Democrats, the divisions between old New Dealers and reformers were taking on an increasingly "ideological" dimension.

American Party Factionalism After the New Deal

In the mid-1960s, at the end of the New Deal era, both parties were moving toward a more ideological form of intraparty conflict. The Republicans had clearly divided into conservative and moderate factions, and the Democrats, into liberal, regular, and southern factions (see Table 1.1). Bifactionalism remains the predominant line of conflict in the Republican party, and the Democratic party has splintered further until it represents a multifactional pattern of intraparty competition.

In the Republican party the factional pattern established in the 1940s between "moderates" and "conservatives" has been intensified and maintained. What has changed is the factional balance between the two factions and the nature of the issues that divide them. As a result of the post–New Deal electoral alignment that brought the southern Republican party to life and transformed the southwestern states' Republican parties (particularly California's) from being largely moderate to largely conservative, the conservative faction consolidated its domination of the congressional party (especially in the House) and completely routed the moderates in recent presidential politics.

Table 1.1. The Pattern of American Major Party Factionalism, 1968 and 1972

	Democrats	Republicans
1968	Regulars Liberals The white South	Conservatives Moderates
1972	New Left/minorities Regulars Neoliberals The South	Social conservatives Libertarians

The Republican right was assisted by the disintegration of the traditional party organizational structure in many states and the advent of the
new populist politics of the issue activists.[25] Moderate Republicans
remain significant only in the U.S. Senate, reflecting the exigencies of
Republican statewide electoral success in certain (generally Democratic-
dominated and culturally liberal) states and among the party's state
governors (an office that does not attract ideological conservatives).[26]
The weakness of the self-described moderates in the current Republican
coalition means that the terms *moderates* and *conservatives* are no longer
accurate with regard to Republican party factionalism. What is emerging
is a new factional cleavage between social and libertarian conservatives
on issues pertaining to civil liberties, sexual morality, law and order,
abortion, and, possibly, free trade.

The political upheaval engendered by the Vietnam War, the civil rights
revolution, and urban violence during the late 1960s divided the Democratic party into three distinct factions: the New Deal regulars, the "new
politics" liberals (to whom I refer as "liberals"), and the white South.

The regular faction—based on the remaining urban political organizations and traditional white ethnic communities—has lost influence in the
party since 1968, as its traditional electoral constituencies either have
been diminished by social and demographic change or have been realigned electorally with the Republican party. Organized labor has been
associated closely with the regular faction since the New Deal, although
labor has since moved toward a less conservative position as minorities,
women, white-collar workers, and public service employees have become larger components of union membership.

The liberal faction, which finally won the national Democratic party
away from the old New Dealers and white Southerners in 1968–72, has
since splintered into a New Left/minorities faction that emphasizes the
demands of "disadvantaged" groups—such as blacks, Latinos, feminists,
and gays—and a neoliberal faction based on the aging baby boomers of
the suburbs and the academy. On social and cultural issues, members of
the latter faction generally adhere to liberal positions, and on foreign
policy issues, they remain hesitant regarding U.S. intervention abroad,
but they have become increasingly antipathetic to traditional Democratic
policies of higher personal taxation and government intervention in the
economy.

The position of the southern faction is anomalous. It has been greatly
weakened in the party by the advent of southern Republicanism and the

post-civil rights movement's electoral realignment in the South, but it remains a vital element of the Democratic national coalition. Unlike the other party factions, the southern faction retains a sectional rather than an ideological orientation, reflecting the persistent (though diminished) regional distinctiveness of the South. Although the southern Democrats retain a balance-of-power position in the Congress, in the presidential party they have been generally weaker, as the other factions have united against southern candidates, except in 1976 and 1992. The southerners have tended to ally with the neoliberals on economics and with the regulars on social and cultural issues and foreign policy. On civil rights the southern faction is no longer distinct from the liberal mainstream of the Democratic party.

This, then, is the pattern of modern American party factionalism, with one party divided into at least four factions and the other into two. In the Democratic case, the factional balance differs at different levels of electoral competition. Since the New Deal the pattern for American party factions has been to move from a clientelistic to an ideological basis. The greater social and economic homogeneity of the country has also led to a marked decrease in the sectional aspects of American party factionalism. Indeed, contemporary party factionalism appears to be based more on occupational or cultural lines, which increasingly transcend state and regional boundaries. The factions have also become institutionalized in a way that they never were before the New Deal, with a high degree of self-consciousness and continuity.[27]

The major exception to the trend toward "ideologization" is the persistence of a sectional southern faction in the Democratic party that cannot be easily assimilated with the surviving regulars into a general conservative Democratic faction. This faction, although probably less cohesive and disciplined than it was during the era of the New Deal, has not disappeared, even with the advent of a competitive Republican party in the South, and its support remains crucial to the Democrats' national electoral success.

The Republicans: Conservative Dominance

Conservative and liberal differences in the Republican party were traditionally based on reactions to the New Deal and foreign policy. Conservative Republicans such as Senator Robert Taft believed that state intervention had to be limited, and he intensely disliked the New Deal system of government that had departed from free-market, individualistic, Republican principles. Moderate Republicans, reflecting their eastern big business and financial base, easily accommodated the popular appeal of the New Deal's social welfare measures. The divisions were even more apparent on foreign policy, in regard to which the conservatives tended toward an isolationist position and the moderates were enthusiastic interventionists and supporters of NATO and the Atlantic Alliance. McCarthyism and anticommunism succeeded in converting the conservative Republicans to a global foreign policy, but they remained more likely to look toward Asia rather than Europe as an American sphere of influence and to want to rely on nuclear and air power rather than committing U.S. ground forces overseas.[28]

Before the civil rights revolution and the revolution in public morality during the 1960s, both liberal and conservative Republicans were generally supportive of liberal positions on civil rights issues and issues dealing with sexual mores such as abortion and contraception. During the 1960s, however, a large section of the GOP also turned ultraconservative on cultural issues as well. Republicans began to doubt the wisdom of civil rights measures when those measures began to focus less on political liberties and more on economic demands that implied some costs for the GOP's white, middle-class, business constituency. Eagerness to cultivate the so-called white backlash vote after the 1960s urban race riots also played a part in the willingness of the GOP's conservative wing to abandon civil rights.[29] Republican moderates refused to move so far in this direction and continued to show concern over the evaporation of the party's Lincolnian heritage.

Between 1960 and 1980 the Republican party moved from a pattern of fairly even competition between its moderate and conservative factions to complete domination by the conservative faction. Indeed, during the 1980s it was highly questionable whether the moderate Republicans were sufficiently viable to be called a faction in any real sense of the word. Whereas conservative Republicans had always been dominant in the congressional Republican party after 1932, in presidential politics the moderates' advantages of money and personnel were usually decisive in the 1940–60 period. In closely contested nominations, the moderates always had more resources than their conservative opponents did to win over doubting delegates.[30]

This situation changed for several reasons. The civil rights revolution and industrial development created a viable Republican party in the southern states that was strongly conservative. New economic centers of power emerged in the Sunbelt states stretching from Atlanta and Miami to Los Angeles, which eroded Wall Street's dominance at the elite level of the party. The Republican right also benefited from electoral shifts by conservative Democrats in the South and the movement of the nominating process toward primary elections and media campaigns that gave inherent advantages to the ideologically committed over the centrists or moderates.

All these factors contributed to the surprising victory of the conservative Barry Goldwater at the 1964 Republican convention, and the moderate wing of the party has never recovered from that blow.[31] Conservative supremacy was reinforced after 1968 by the move toward an even greater role for primary elections in the presidential candidate selection process, and Nelson Rockefeller's campaign in that year was the last serious moderate Republican bid for the party's nomination.

The picture in Congress is slightly different, though here one must be careful to differentiate between the two chambers. In the House the moderate Republican element was reduced by redistricting (the concentration of the Republicans into extremely homogeneous and conservative districts) and realignment to a small remnant around the Wednesday Group. Ambitious Republican moderates had few incentives to stay in the House for an extended period as a minority faction in an almost-permanent minority party, and so they tended to use the chamber as a stepping-stone to the Senate or a statewide office. In the more closely contested Senate, the moderates fared better because of the greater social and ethnic heterogeneity of the states as opposed to that of the congressional districts and by the Republicans' realistic aspirations toward controlling that chamber. To win in many states—in New England, the Upper Midwest, and the Pacific Northwest—Republicans needed to field moderate candidates attuned to their political cultures.[32]

In statewide and local elections focusing on administrative capacity rather than ideology, the moderate Republicans fared better still, electing a sweep of governors in 1986 and 1990. The constituent-service orientation of most state legislatures is also a moderating force in the GOP, although here moderates suffer from some of the same factors that have eroded their numbers in the U.S. House: endemic minority status and hostile redistricting.

As the moderate wing has been marginalized at the national level, even moderates in Congress have been forced to pay at least lip service to conservative ideals. There are now no genuine economic or foreign policy liberals in the Republican party at any level, and moderation tends to manifest itself on civil rights and civil liberties issues. This is a possible wedge by which a moderate wing could revive itself by dividing the conservative faction, as a substantial number of libertarian and upper middle-class suburban Republicans are deeply uncomfortable with the religious right's positions on abortion and other matters of sexual conduct. Although clear-cut factions on social conservative versus libertarian lines have not yet taken shape, the insurgent presidential candidacy of Patrick Buchanan in 1992, which emphasized social conservatism, protectionism, and an isolationist foreign policy, and the widespread dissatisfaction with the national GOP's strong antiabortion policy among socially moderate and libertarian Republicans may be the harbingers of future factional divisions within the GOP during the 1990s.

Democratic Party Factionalism in the 1990s: A Complex Pattern

Within the contemporary Democratic party it is possible to isolate four factions: the New Left/minorities faction, the regulars, the neoliberals, and the South.

The New Left/minorities faction generally supports interventionist government (particularly to promote the interests of blacks, Latinos, gays, and women). It also has an endemic distrust of American foreign and defense policy, particularly with regard to the Western Hemisphere and more generally with the developing world. Finally, adherents of this faction tend to be extreme modernists in cultural terms, being highly suspicious of all traditional cultures and values and appeals to such values and being vehement in their defense of the rights that they feel are threatened by traditional values. The New Left/minorities faction has its electoral base among minority, feminist, and gay activists, on college campuses, and among public-sector employees. Their most prominent leader in recent years has been Rev. Jesse Jackson.

The regulars are those Democrats who still adhere to the fundamental premises of Roosevelt's New Deal coalition emphasizing government intervention to help the disadvantaged, a strong defense policy, and respect for traditional values in culture. This faction dominated the presidential Democratic party until the late 1960s and has been greatly weakened at all levels of the party since. The reasons are that its social base—labor unions and the blue-collar, white working class outside the South—has shrunk in relative terms and has also been wooed with some success by the Republicans in recent presidential elections.

The regular faction remains based largely in the older industrial states and in urban areas, among Jewish and Roman Catholic, white, ethnic, middle- and working-class Democrats. Although not antagonistic toward civil rights for women and minorities, this section of the Democratic party has greater respect for the traditional values espoused by their working-class constituents. The traditionalists also do not share the New Left's fundamental ambivalence about the rectitude of American values and American patriotism. In recent years the regular wing of the party has become extremely receptive to protectionist sentiments because its manufacturing base in traditional industries is threatened by Japanese and European competition. Recent leaders of the regular faction in the Democratic party have been former Vice-President Walter Mondale, Senator Daniel Patrick Moynihan, House Speakers Tip O'Neill and Tom Foley, and House Ways and Means chair Congressman Daniel Rostenkowski.[33]

The neoliberal faction also has its origins in the 1968–72 revolt against the Democratic regulars, but unlike their former allies in the New Left/minorities segment of the party, since 1972 the neoliberals have moved increasingly toward the center of the political spectrum. Indeed, many contemporary neoliberals—particularly Gary Hart and Paul Tsongas—have broken with traditional Democratic economic doctrine on social welfare, Keynesian economics, and redistributive taxation, although they still believe that government has a critical role to play in promoting economic growth. Neoliberals have sought to play down ideology and prefer to focus on a technocratic approach to government, as epitomized in 1988 presidential candidate Michael Dukakis's phrase "This election is not about ideology, it's about competence."

In contrast with the other factions of the party, the neoliberals remain ardent free traders. They also generally remain strongly supportive of New Left positions on civil liberties and the rights of women and minorities. On foreign policy the neoliberals also still tend more toward the dovish side of the party. Their base of electoral support has been in the educated, suburban, professional middle class. Prominent neoliberals in recent years have included former senators and presidential candidates Gary Hart and Paul Tsongas, and 1988 Democratic nominee Michael Dukakis. President Bill Clinton also appears to lean most toward this section of the party.

Across the board, the southern faction is generally more conservative than the other factions. Except for the textile-producing areas, it generally supports free markets and free trade but also generous federal government assistance for oil and agriculture. It generally opposes labor unions and government intervention in the economy except in areas important to the South, such as agriculture. On cultural issues this faction reflects the cultural conservatism of the southern region, and in addition, southerners are more likely than Democrats elsewhere to take the conservative position on issues of religion, civil liberties, and personal morality. On foreign and defense policy the southern faction also tends toward a conservative position, partly because of the large number of military installations and defence industries in the South.

On civil rights, as the southern Democratic party has come to rely heavily on black voters and much of the hard-line segregationist white vote has shifted to the Republicans, the southern Democrats have turned around 180 degrees to a very strong pro–civil rights position.[34] Thus on economic issues the southern Democrats tend to be closer to the neoliberals, but on cultural and foreign policy matters they are generally to the right of the other Democratic factions. The electoral base of the party's southern wing lies among southern blacks and southern white working-class and rural voters. Its most prominent leaders in recent years have been Senators Sam Nunn, Lloyd Bentsen, and Charles Robb.

The factional balance varies at different levels of the party. In presidential politics the regular faction lost control to the New Left in 1972, but after the McGovern debacle, the New Left has been unable to nominate another presidential candidate. With the exception of the regular Walter Mondale in 1984, the Democratic presidential candidates since 1972 – including the ostensible southerner Jimmy Carter – have come from the neoliberal section of the party: more conservative than the regulars on economics, liberal on social and cultural issues, and ambivalent regarding U.S. intervention abroad.

The move toward a system of presidential nominations through primary elections rather than in national conventions has enormously benefited the neoliberal and New Left factions, at the expense of the southerners and the regulars. In primary elections, which are determined by the ability of the candidates to mobilize middle-class, single-issue, and ideological activists, neoliberal and New Left candidates have inherent advantages. White-collar suburbanites, teachers, and militant activists are much more likely to get involved in campaigns and to show up at the polls, and in low-turnout primaries this advantage is decisive.[35] The crucial influence of the New Hampshire primary in the process particularly helps the neoliberals. Despite its conservative reputation, New Hampshire has consistently launched neoliberal Democratic presidential candidacies, which have converted the New Hampshire momentum into eventual nomination (Carter, Dukakis, Clinton) or near nomination (Hart). The urban, industrial states that constitute the geographic base of the New Left and the regulars, as well as the southern states, have played a much smaller role in this process.[36]

At the congressional level, which is less focused on ideology and more on issues of constituency representation and delivery of services to states and districts, the regulars and the southerners fare better, although the influence of both has eroded greatly since the mid-1960s when the southern Democrats controlled the Congress through the committee system and the seniority rule.[37] The southern Democrats nevertheless remain a crucial voting bloc in Congress despite their reduced numbers, and they have been consistently able to hold districts and states that have decisively rejected Democratic presidential candidates over the past quarter-century.

In state and local politics, which is even less ideologically focused than that of the Congress, we find a similar pattern: slowly growing neoliberal and New Left dominance over regular Democrats in most northern industrial states, and continuing conservative dominance in southern state and local politics, with a growing neoliberal presence.

In short, the Democratic party displays a complex but consistent factional pattern in both geographic and class terms. However, there are clear indications of a long-term erosion in the influence of the regulars at the party's elite level and a much slower erosion of the position of the southern Democrats. At all levels, but particularly in presidential politics, the neoliberal faction is coming to dominate the Democratic party as the conservatives came to dominate the GOP. The New Left faction, with its base among minority voters and the militant issue activists, is an important presence in presidential politics, but failing an alliance with the neoliberals, it cannot win the presidential nomination in the face of the opposition of the other three factions. On Capitol Hill and in the states, the New Left has little influence.

Implications

The transformation of American party politics from a system based on strong regional and ethnic loyalties to one based on ideologies has had a profound effect on the nature of American party factionalism. Party factions traditionally based on regional–ethnic ties have been replaced by factions based on ideology. American party factions remain relatively loose and undisciplined in comparison with those in other western democracies (which have parliamentary systems of government and much stronger party loyalties), but there has been an intensified pattern of factional cleavage in each party in recent years.

Indeed, in both parties we can see signs of the "institutionalization" of the factions, with the formation of permanent institutions representing the factional viewpoint aside from particular presidential candidacies. Southerners and regulars in the Democratic party formed the Democratic Leadership Council (DLC) to promote a "moderate" viewpoint.[38] The New Left formed the Coalition for Democratic Values (CDV) to promote its ends.[39] The conservative faction in the Republican party has built an entire infrastructure of interest groups, publications, and think tanks to promote conservative views. Moderates also counterorganized during the 1960s through the Ripon Society, although this has become a much less significant force in the party in recent years as moderate influence has further diminished.[40]

Factional institutionalization has also taken place in Congress. Liberal Democrats have been organized around the Democratic Study Group (DSG) since the late 1950s, and during the 1980s southern Democrats met regularly and wielded their legislative influence through the Conservative Democratic Forum (CDF). On the Republican side the equivalent of the DSG for conservative Republicans was originally the Republican Study Committee (RSC), and moderates organized through the Wednesday Group. In recent years as the Republican Study Committee came to include too many House Republicans to be an effective factional force, younger conservatives have utilized the Conservative Opportunity Society (COS) to promote their views in the House.

Another interesting difference between the pattern of factional politics in the United States in the pre–New Deal era and in the modern period is the means of mediating factional conflict. In the pre–New Deal period and into the 1960s, intraparty factional conflicts based on local or sectional claims were mediated through bargaining between party leaders at the national party convention, or through the congressional committee system and the seniority rule. One of the weaknesses of the current American party system is the absence of effective means of mediating factional conflict, since the power of the mediators – the party leaders and elected officials – has been greatly reduced by the move toward primary elections and the disintegration of traditional party organizations. As mentioned earlier, the new playing field favors the intensely committed and the ideologue, and it gives candidates every incentive to polarize factional debate in order to differentiate themselves and get resources, media attention, and activist support. The consequence is bitterly contested nominations that divide the party as a whole and preclude the kind of postconvention coalition building that the old politics, for all its multifarious shortcomings, permitted.

The weakness or absence of mediating institutions has been a particular problem for the Democratic party, which has recently endured more factional strife than has the conservative-dominated GOP. These factional struggles have seriously damaged the prospects of recent Democratic presidential candidates, all of whom have been bruised by fierce nomination contests. At the congressional level the erosion of the committee system and the greater importance of the party caucuses and floor debate in both Houses have contributed to the general decline in comity on Capitol Hill and in the effectiveness of Congress as an institution.

Thus American party factionalism during the 1980s forced American politics into an ideological straitjacket that did not reflect the positions of either most voters or most active politicians. It did so because the dynamics of presidential nominating politics and – although to a lesser extent – congressional politics gave inherent advantages to the more ideologically committed factions, the Democratic New Left and neoliberals and the GOP conservatives. The fact that the former dominated the Democratic presidential party while the latter dominated the Republican congressional party made it almost impossible for the parties to control those branches of the federal government. Byron Shafer pointed out that neither party appeared likely to win unified control of the federal government during the 1980s, since neither of the dominant ideological factions was prepared to make the doctrinal compromises necessary to enable the Republicans to win Congress or the Democrats to win the White House: "The activist base of each political party had become effectively committed – energetically, consciously, even desperately committed—to precisely those positions on which their party would necessarily lose one or another major public office."[41] The rise of ideologically committed factions in American politics thus contributed, for better or worse, to the entrenchment of divided government in the United States during the 1980s.

2

The Old Southern Democracy and Its Erosion, 1876 to 1965

No southerner ever dreams of heaven, or pictures his Utopia
on earth, without providing room for the Democratic party.
John Crowe Ransom[1]

Most of our present-day conceptions of southern politics remain grounded
in the era of the "Solid South," the period between the end of Reconstruction and the Voting Rights act of 1965. During this time the Democratic
party achieved such a degree of political dominance over the entire region
that it became identified by both its followers and its opponents as the
"party of the South" above all else. The virtual one-party system established by the Democrats in the former Confederate states, however, was
not a dictatorial system in the sense of rule by a monolithic entity. Solidity
behind the Democratic label was necessary for national politics, but that
label covered a variety of political phenomena—factions, demagogues,
violence, racism, populism, authoritarianism—that were truly unique to
the South and seemed bizarre to nonsoutherners.

This chapter sketches the origins and development of the Solid South,
the operation of the system at its height, and the factors that led ultimately
to its demise. The story has been told frequently and in much greater
detail elsewhere, but it is reiterated here as a reminder of the nature of the
pre–civil rights southern Democratic party in national politics relative to
the southern Democratic party that succeeded it. We begin by retreating
further back to the antebellum South to understand the nature of southern
political culture and the reasons for the ascendancy of the Democratic
party after the Civil War.

27

Politics in the Antebellum South: Honor and Slavery

From the earliest settlements in the southern states, it was clear that those people who had been attracted there came from a culture and a social milieu different from those of the people that had landed in New England. The Tidewater South was settled largely by lesser gentry and their indentured servants from southern and western England, who supported the Anglican church and the Cavaliers during the English civil war. By contrast, the Puritan settlers of New England tended to be artisans or independent yeomen from East Anglia, who supported Calvinist churches and Oliver Cromwell (another East Anglian farmer) against the Stuarts.[2] In the southern states a plantation economy developed around cash crops such as tobacco, sugar, and cotton which, together with the more hierarchical cultural background of the southern migrants, produced a society that evolved in a quite different direction from that of the northern colonies. Whereas in the North, wealth was based in urban centers and manufacturing, in the South, wealth and economic power lay not in the towns but in the farms and plantations. Revolutionary Massachusetts was dominated by urban radicals, but the southern leaders of the revolutionary war were gentleman farmers and slaveholders.[3]

There was, however, more than one stream of migration into the South that also supplied some of the more radical southern leadership in the revolutionary war. In the late eighteenth century, the backcountry mountain South was settled by large migrations of Scotch–Irish, whose cultural and political mores differed from those of the Tidewater. Scotch–Irish society was ferociously Calvinist, individualist, egalitarian, and clannish, and their economic system—based on small-scale subsistence farming on the frontier—did not require slave labor to the extent that the plantation economy of the Tidewater did.[4]

Thus the antebellum South was an amalgam of two political cultures, and the emergent southern political culture took on aspects of both: From the Tidewater came the concern for manners, deference, and the notion of the "gentleman"; from the backcountry came the spirit of individualism, populism, fundamentalist religion, and fierce loyalty to one's "kind."[5] W. J. Cash has provided the most vivid depiction of the antebellum southern "mind" and the southerner's conflicting hedonistic and religious impulses:

> To stand on his head in a bar, to toss down a pint of raw whiskey at a gulp, to fiddle and dance all night, to bite off the nose or gouge out the eye of a

favorite enemy. To fight harder and love harder than the next man, to be known eventually far and wide as a hell of a fellow—such would be his focus.[6]

What our Southerner required, on the other hand, was a faith as simple and emotional as himself. A faith to draw men together in hordes to terrify them with Apocalyptic rhetoric, to cast them into the pit, rescue them, and at last bring them shouting into the fold of Grace.[7]

According to Bertram Wyatt-Brown, both cultures shared a devotion to the concept of honor and the importance of saving face. Indeed, notions of honor defined southern society almost to a greater extent than did slavery:

> Above all else, white southerners adhered to a moral code that may be summarized as the rule of honor . . . The sources of that ethic lay deep in mythology, literature, history, and civilization. It long preceded the slave system in America. Since the earliest times, honor was inseparable from hierarchy and entitlement, defense of family blood and community needs. All these exigencies required the rejection of the lowly, the alien, and the shamed. Such unhappy creatures belonged outside the circle of honor. Fate had so decreed.[8]

Interestingly, as the cotton crop and the slave system began to spread westward, into the Piedmont (central North Carolina, northern South Carolina, and Georgia) or backcountry areas like Mississippi and east Texas, the residents there also gained a stake in the slave economy, but their brethren in the high Appalachian mountain areas (eastern Tennessee, western Virginia, and North Carolina) had no reason to identify with Tidewater slavery and later repudiated the political leadership of the slave-owning class.

The antebellum South was not a social monolith, and neither was it politically monolithic, as political divisions reflected the economic and cultural cleavages in southern society. The Democratic party founded by Andrew Jackson, a Scotch-Irishman from Tennessee, was strong in the backcountry and the frontier, where its economic populism and egalitarian rhetoric struck a chord with the local population. The Whigs, by contrast, were stronger among the cotton planters and merchants of the Tidewater and the "black belts."[9]

The rise of the slave issue to the forefront of national debate during the 1850s destroyed the balanced two-party competition that characterized the antebellum South, and it united both the Tidewater and the back-country against the abolitionist sentiment that swept the North after the Kansas–Nebraska Act of 1854.[10] Apart from the remote mountain sections of the backcountry where slavery had never penetrated, the South as a regional entity—almost a nation—was forged in the political struggle over slavery and the Civil War that followed. In the process, the Whig party was destroyed, and the southern Whigs repudiated by the Republican party in the North had no alternative but the Democrats, who, under Stephen Douglas, spent the 1850s striving for continued compromises that would avert the coming catastrophe.

Reconstruction and Restoration: The Forging of the Solid South, 1860 to 1896

In the last years of the nineteenth century, the slave-owning political elites who led the southern states into secession from the Union and into civil war were able to use the experience of war, defeat, and reconstruction to form a southern social and political system on their own terms. Southern whites, whether rich or poor, backcountry or Tidewater, were completely alienated from the "party of Lincoln" as a result of their experiences in the war and their resentment at the attempted Republican Reconstruction in the South between 1865 and 1876. These resentments were also fueled by fears about the economic and political power of the freedmen. Middle-income and poorer whites feared losing their economic and social position to the freed blacks, who might also hold the electoral balance of power in most of the southern states. Thus by the end of Reconstruction in 1876, the white South (outside the "mountain Republican" redoubts) was already effectively united in hostility toward the Republican regime in Washington.[11]

With the withdrawal of the federal troops and the restoration of the Tidewater planter/merchant elite to political power after 1876, the one-party Democratic South began to take shape. Its economic outline was already clear. Despite significant industrial development in some areas after the Civil War (particularly in textiles and lumber), the South remained a predominantly agricultural economy based on the three staple plantation crops: tobacco, sugar, and cotton. The planter elite's economic power was reinforced by the development of tenant farming and the crop lien and sharecropping systems, all of which maintained black and poor white tenant farmers in a state of abject economic dependence on the so-called bourbon elite of planters and their banker and merchant allies, in the South's few sizable urban centers.[12]

This agrarian, almost feudal, economic system was accompanied by a rigid system of racial segregation. The Jim Crow laws introduced after 1876 were intended to keep blacks and whites separated in all aspects of social life, including education, transport, and public accommodations. After the U.S. Supreme Court had upheld the constitutionality of seg-regation its 1896 *Plessey v. Ferguson* decision, strict separation of the races became almost universal in the southern and border states.[13]

The postbellum South's economic and social system was politically protected by an electoral system that kept political power in the hands of a very small cadre of upper-status whites. The Democratic label was the chosen mechanism by which this domination was reinforced. After the restoration, the Democrats returned to political dominance in most southern states as the southern white electorate reacted against the Republican attempts at Reconstruction. However, black and mountain white votes did keep the Republican party alive in much of the South during this decade.[14]

Although the Republican minority could generally be contained, a more serious threat to "bourbon" control over southern politics emerged with the populist movement in the early 1890s. Populism's appeal to poorer framers and its (initial) biracial tone gave the South a brief glimpse of the emergence (or reemergence) of a southern political alignment based on economic class cleavages rather than race and the Civil War. The problem that the populists faced, however, was that the poorer white constituency that they represented was also extremely vulnerable to fears of social and economic subordination to their black counterparts.[15]

During the 1890s the South's planter elite played ruthlessly and re-lentlessly on the anxieties of the lower-status whites. In so doing they not only defused the populist threat but also were able to reinforce their own political position. Disguising their real intentions in the progressive/ reformist "good government" rhetoric of the day, all the southern states implemented a series of measures that effectively excluded not only virtually the entire black population but also most of the poorer whites from participation in politics. New state constitutions introduced the devices intended to keep the political process limited to upper-status whites: the poll tax, the literacy test, and the grandfather clauses. If all these failed, then outright intimidation could always be employed as a last resort.[16]

With its black electoral base barred from the polls, the Republican party was virtually driven into extinction as an electoral competitor, and it survived largely on sufferance merely as a sham-patronage organization. Behind the Democratic party label, a very narrow, white, upper-status elite controlled politics in every southern state. In order to provide a modicum of electoral competition, the primary election was introduced (beginning in Florida in 1900), along with provisions for a "runoff" election between the top two candidates if no candidate should secure an outright majority. This was in case an "undesirable" contender was able to win a plurality victory in the first primary, with the vote splintered among several candidates. By defining the Democratic party as a private association, southern Democratic leaders were also legally entitled to limit participation in party primaries to white voters only, and the so-called white primary was probably the most effective of all the means of disfranchisement employed in the south after 1896.[17]

The takeover of southern politics by the bourbon elites of the black belts and the deliberate suppression of the Republican party and the populists met with little resistance from outside the South. The nationally dominant Republican party did not need the South electorally after 1896, and with the experience of Reconstruction still fresh in their minds, Republican leaders were too preoccupied with building up the United States' industrial base in the North to divert resources to another confrontation with the white South. The Democratic party outside the South consisted largely of urban, ethnic, immigrant, political organizations, which although they had nothing culturally in common with the southerners, shared the latter's resentment of the "Yankees," the northeastern, white, Anglo-Saxon, Protestant ruling class. They held no great political or social affection for the disfranchised southern blacks and were thus unlikely to do anything to undermine their party's grip on its most reliable region in national elections.

The price they paid, however, was that no genuine white southern politician could aspire to the presidency in either party. Memories of the great conflict were still too fresh to permit a "Reb" or the son of a Reb to sit in the White House. Yet as we shall see later, the southern Democrats did exert considerable influence over national politics during the Solid South era, even if they were implicitly excluded from obtaining the highest office.

The Solid South at High Tide, 1896 to 1948

The Solid South at high tide was the political South described by V. O. Key, Jr., in his *Southern Politics*, and the foundations of the Solid South were laid in the 1876–96 period. The devices introduced in the 1890s—the poll tax, the literacy test, the white primary, the runoff, the grandfather clause—virtually eliminated black voting and excluded a substantial number of poorer whites.[18] In the circumstances of the 1920s when voting participation rates in the southern states sank to 20 percent or less, this left a tiny electorate that was easily controlled by the planter/ merchant elite of the black-belt counties. To reinforce their dominance, electoral districts for the U.S. House and southern state legislatures were ruthlessly gerrymandered in favor of rural and small-town areas and against the urban centers. Disfranchisement also established single-party rule by the Democrats and eliminated the Republican party as a serious political force in most of the Southern states.

The rationale for the system was, of course, the maintenance of strict racial segregation and the complete social and economic subordination of the blacks. Lower-status whites were also largely politically and eco- nomically powerless, but their loyalty to the system was ensured by their fears of losing social and economic status to the blacks if the system were dismantled. Southern elites manipulated these anxieties to establish and perpetuate a political system that operated very much according to their interests and on their terms. The system operated most virulently in the black-belt cotton counties that held the largest plantations and where the black population was highest. Key argued that the plantation owners of the black belt used segregation, disfranchisement, and the one-party system to impose their rule over the entire South, just as they had led the region into the disastrous secession and Civil War of the 1860s:[19]

> The black belts make up only a small part of the area of the South and — depending on how one defines black belt—account for an even smaller part of the white population of the South. Yet if the politics of the South revolves around any single theme, it is that of the role of the black belts. Although the whites of the black belts are few in number, their unity and their political skill have enabled them to run a shoestring into decisive power at critical junctures in southern political history.[20]

The more bizarre features of southern politics could also ultimately be explained by the nature of the system. The stifling of party competition and meaningful elections gave rise to chaotic factional politics in the various southern states. Key contended that in a situation of genuine party competition, the interests of the "have-nots" are better represented because the parties have to strive for the support of all elements of the electorate in order to succeed:

> The grand objective of the haves is obstruction, at least of the haves who take only a short-term view. Organization is not always necessary to obstruct; it is essential however, for the promotion of a sustained program in *[sic]* behalf of the have-nots, although not all party or factional organization is dedicated to that purpose. It follows, if these propositions are correct, that over the long run the have-nots lose in a disorganized politics. They have no mechanism through which to act and their wishes find expression in fitful rebellions led by transient demagogues who gain their confidence but often have neither the technical competence nor the necessary stable base of political power to effectuate a program.[21]

Under the one-party—or effectively no-party—politics of the Solid South, political conflict revolved not around class lines but around personalities and localistic concerns. Representation of the disadvantaged would have entailed a degree of political mobilization and organization that was almost explicitly precluded by the one-party Democratic system. Some states developed a politics of "every man for himself," with no continuity even between the party factions and candidates (Florida). Other states developed a system that had an enduring pattern of conflict between two major factions (Louisiana, Georgia). In Virginia and Tennessee, a single faction largely controlled the politics and government of the state. Rarely, however, did these factional alignments substantially challenge the southern economic and political status quo.[22]

Nevertheless, the system could not entirely suppress class issues all of the time, even in its heyday, and the populist tradition that had been present among lower-status whites in the South since the time of Jackson resurfaced periodically during the Solid South era. In many cases this populism was purely rhetorical and actually operated to reinforce the status quo. Colorful language and rhetoric were obvious means by which a candidate could differentiate himself from the pack of aspirants in the first primary election, and the populist impulse could also be easily diverted in a racist direction to arouse poor whites' anxieties about blacks. The recurrent appearances of characters such as "Cotton Ed" Smith (South Carolina), Eugene Talmadge (Georgia), and Theodore

Bilbo (Mississippi) with their exaggerated rhetoric gave rise to the stereotype of the southern demagogue: racist, menacing, and, to non-southerners, displaying all the most reprehensible features of southern politics and society.

At times genuine populism did nonetheless present a serious challenge to bourbon domination in several southern states. In Texas, James and Miriam Ferguson ("Pa and Ma") espoused populist politics in the 1920s and 1930s and were electorally successful. This tradition was inherited by later Texas politicians such as Sam Rayburn and Lyndon Johnson, who regarded their primary mission in politics as aiding the disadvantaged.[23] The relative weakness of the cotton producers and the low numbers of blacks relative to those of the other southern states also assisted the rise of a more class-oriented type of factional politics in the Lone Star State, even before the New Deal.[24]

The most serious challenge, however, occurred in Louisiana, where Huey Long defeated the political elite through corruption, outright disregard for the law, and sheer ruthlessness and intimidation, which during the 1930s prompted his detractors to describe his Louisiana as a virtual "police state." In fairness to Long, however, it is unlikely that a genuine populist could have succeeded in any other fashion in the political context of the Solid South. Long did implement genuine measures in favor of the disadvantaged in his state, and he did it without frequent recourse to race baiting, in contrast with more demagogic populists. Long's organization survived his own assassination in 1935 and persisted as a major force in Louisiana politics under his brother Earl into the 1960s. Huey Long demonstrated that the system could, on occasion, be beaten, although the price of doing so was the creation of a near-dictatorship in Louisiana, rather than more competitive and responsive two-party politics.[25]

The effects of the Solid South also were manifested in national politics. In fact, the South's "solidity" behind the Democratic party was largely for national political purposes, since southern political elites reckoned that "solidity" was the most effective means to prevent any tampering with the South's peculiar socioeconomic arrangements from the outside.

Between 1876 and 1948 the South was the heart of the Democratic party's presidential coalition. No region was as Democratic as was the South in presidential elections, although because of the national stigma attached to the South after the Civil War, it was all but impossible for a southerner to win the party's nomination.[26] The solidity of the South was achieved by various means in presidential nominating politics, all of which were designed to prevent the national party from ever seriously addressing the race question.

Southern delegations were selected in closed caucuses or by state committees, and they consisted largely of representatives of the white power structure that had been erected in the southern states in the final decades of the nineteenth century. Their essential task at the national convention was to preclude the selection of a nominee "unacceptable to the South" (i.e., unsound on race). To maximize the bargaining power of their states and of the region as a whole, southern convention delegations generally adhered to the so-called unit rule, casting the delegation's total convention vote as a bloc on the floor of the convention (see Table 2.1).[27]

The combined voting power of the southern delegations was thus a formidable barrier to overcome for any candidate regarded as unfavorable to the South. Yet in case such a candidate did appear, the South had an additional device that essentially guaranteed that no Democratic nominee would be selected who might offend southern sensibilities. This was the "two-thirds rule" that prevailed at Democratic conventions up to 1936. Nomination required two-thirds of the total convention vote, and this more or less gave a veto over the nominee to the southern delegations voting in unison. The two-thirds rule also increased the possibility of deadlocked conventions that embittered Democrats and damaged their prospects in the general election (the 103-ballot marathon of 1924 being the most egregious example).[28]

In Congress, southern solidarity behind the Democratic party operated in a similar fashion with regard to issues pertaining to race. During the 1920s the South completely dominated the Democratic caucus in both Houses of Congress, and it controlled the Democratic congressional leaders (see Table 4.1).[29] Their numbers were sufficient to block any unwelcome legislation with regard to the southern socioeconomic system, although given the Republican majorities in Congress for most of this period, the southerners were unable to implement federal agricultural relief that would have benefited their region.[30]

Southern influence in Washington was also being gradually strengthened by the development of the committee seniority system in Congress after the revolt against Speaker Joseph Cannon in 1910. In that year an alliance of Democrats and progressive Republicans stripped the House Speaker of much of his power over committee assignments, chairmanships, and the referral and scheduling of legislation. As a consequence, the party caucuses and the party leadership became much less important

Table 2.1. Southern Influence at Democratic National Conventions, 1876 to 1992

Year	Nominee	Ballot	South's Percent of Total Vote	South's Percent for Winner	South's Percent of Winner's Total
1876	Tilden	2	29	88	35
1880	Hancock	2	29	47	31
1884	Cleveland	2	29	39	20
1896	Bryan	5	28	100	40
1904	Parker	1	28	97	41
1912	Wilson	45	27	48	22
1920	Cox	44	27	52	22
1924	Davis	103	27	75	39
1932	Roosevelt	4	27	100	33
1948	Truman	1	28	18	6
1952	Stevenson	3	28	8	5
1960	Kennedy	1	27	3	2
1968	Humphrey	1	23	84	29
1972	McGovern	1	23	27	11
1976	Carter	1	22	89	26
1980	Carter	1	24	79	30
1984	Mondale	1	27	55	26
1988	Dukakis	1	27	64	25
1992	Clinton	1	28	95	33

Note: Table shows conventions where the nomination was seriously contested. The South includes Oklahoma and Kentucky.

Sources: Gerald Pomper et al., *The Election of 1984: Reports and Interpretations* (Chatham, NJ: Chatham House, 1984); Pomper et al., *The Election of 1988* (Chatham, NJ: Chatham House, 1989); Pomper et al., *The Election of 1992* (Chatham, NJ: Chatham House, 1993); *Congressional Quarterly Almanac 1976* (Washington, DC: Congressional Quarterly Press, 1977); *Congressional Quarterly Almanac 1980* (Washington, DC: Congressional Quarterly Press, 1981); and *National Party Conventions 1831–1972* (Washington, DC: Congressional Quarterly Press, 1976).

in Congress, and in their place the congressional committees became more important. Committee assignments were then based on the interests of the districts or states of the members. The committees also became independent from the party leadership, with chairmanships and ranking memberships being awarded on the basis of seniority. Members also tended to defer to the wishes of committees (i.e., reciprocity) on the floor of each chamber. All of the most important congressional business was now carried out in committee.[31]

This system greatly benefited the South, of course, because the one-party system meant that southerners were likely to accumulate more seniority than did other members. When the Republicans held majorities in Congress, the full effect of this did not become apparent, since the crucial chairmanships were not in Democratic hands. In the era of Democratic dominance that began in the 1930s, however, the South established an iron grip on Congress through the committee seniority system. During the 1950s William S. White wrote of the U.S. Senate:

> The place is, to most peculiar degree, a *Southern* Institution engrafted upon, or growing in at the heart of, this ostensibly national assembly of the sages.
>
> So marked and so constant is this high degree of Southern dominion, in spirit or in fact in the varying times, that the Senate might be described as the South's unending revenge upon the North for Gettysburg.[32]

Needless to say, southern solidarity on Capitol Hill on any issue that impinged on the "southern" social and political system was absolute.[33]

For most of the 1896–1932 period, the Republicans offered no serious challenge to the solidly Democratic South. There were two obvious reasons for this. First, the Republicans did not need southern support to win presidential elections or to control Congress, so they did not waste their electoral resources there. Second, and perhaps more important, the Republicans believed in southern solidity every bit as much as the Democrats did and did not feel that they could ever overcome the region's endemic hostility toward the party of Lincoln and Reconstruction. Periodic outbreaks of Republicanism in the South even during this period may give us some grounds for skepticism in this regard.[34]

Before disfranchisement in the 1890s, the GOP was still obtaining at least a quarter of the statewide vote in most southern states, and even after the establishment of the "Solid South," the Republicans retained a substantial presence in the mountain south—eastern Tennessee, western North Carolina, and scattered counties in Virginia, Alabama, Texas, and Arkansas.[35] In 1928 the most serious crack in the Solid South since its inception opened when Republican Herbert Hoover carried all the "Rim South" states—Tennessee, Virginia, North Carolina, Texas, and Florida—and made a respectable showing in some of the others. Only those states and counties with the largest black populations, the lowest levels of urbanization, and the greatest dependence on the plantation economy stayed loyal to Democrat Al Smith. This election has been dismissed as an aberration caused by the adverse reaction to Smith's Catholic faith in the southern Bible Belt, but it is clear that the pattern of Hoover's vote did foreshadow the later evolution of presidential Republicanism in the South in the 1950s. The Republicans did not appear eager to exploit this opening, however, and the advent of the economic holocaust of the Great Depression retarded their progress in the South by at least a generation.[36]

Thus by the early 1930s, the southern plantation elite had established a one-party political system that largely succeeded in suppressing challenges to their power inside the southern states, by the effective exclusion of "class issues" from political debate and the exclusion of the "have-nots" (black and white) from participation in the system. At the national level, solidarity led to effective southern domination of the Democratic party and to the preemption of challenges to the southern political system from that quarter.

The southern influence over the national Democratic party persisted into the New Deal era. Although the Roosevelt administration posed no threat to the southern system of segregation, Roosevelt's measures to relieve the Depression, particularly agricultural relief, actually reinforced Democratic strength in the white South during the 1930s.[37] The economically backward South benefited more than did any other region from the New Deal's social and agricultural relief programs and public works projects such as the Tennessee Valley Authority. Moreover, the establishment of huge Democratic majorities in Congress in combination with the seniority rule gave the lion's share of the all-important committee chairmanships to southern Democrats.

The advent of the southern domination of Congress effectively stifled any congressional or executive branch attempts to attack the Jim Crow system. In fact, Roosevelt found it impossible to get even fairly mild "antilynching" bills passed at the height of his political powers, and he was too canny a politician to risk his southern Democratic base by taking up the cause of black civil rights, despite pressure from some other New Dealers such as Eleanor Roosevelt and Harold Ickes.[38]

When the New Deal became too "radical" for the liking of the southern political establishment—particularly with reference to Roosevelt's "court-packing" plan in 1937—the southern Democrats formed the "conservative coalition" with the minority Republicans, which effectively controlled the Congress up to the mid-1960s.[39] This coalition also ensured that the power of the southern committee barons would be sustained on the floor of both chambers. Although some elements in the "class-based" Democratic majority outside the South began to chafe at the restrictions that appeasement of the southern Democrats placed on the expansion of the

New Deal, the South generally conceded enough ground on issues such as legislation guaranteeing the rights of labor unions to prevent the breakup of the New Deal coalition. In essence, the South supported New Deal programs to benefit the northern working class as long as the representatives of the latter continued to turn a blind eye to the southern social and economic system. Richard Bensel described the bargain that lay at the center of the New Deal Democratic coalition:

> During the two decades following the war, the conditions for a bipolar New Deal coalition were tolerance of race segregation in the southern periphery, the relative isolation of class conflict in the industrial core, and a dependence on administrative discretion and legislative decentralization as methods of minimizing intraparty political conflict. Before 1930, the Republican party had continuously exploited sectional conflict as a strategy for retaining national power; in the next thirty-five years, the Democrats sought to minimize regional competition to the same end.[40]

Nevertheless, despite the evident reinforcement of Democratic party dominance over southern politics that took place during the 1930s, the New Deal was already sowing the seeds of the eventual destruction of the Solid Democratic South.

The Civil Rights Movement and the Disintegration of the One-Party South, 1948 to 1965

The collapse of the Solid South during the post–New Deal period was due to social and demographic changes within the Democratic party's national coalition and within the South itself that destroyed the social and economic environment that had sustained the one-party Democratic political system. Despite Roosevelt's success in avoiding the issue, the logic of the New Deal realignment meant that the Democrats as a national party would ultimately have to address the problem of the southern race system. During the 1930s, blacks in the northern urban centers – where they could vote – had realigned in favor of the Democrats, in support of Roosevelt's relief programs.[41] As the GOP began to recover its electoral strength in areas such as the Midwest in the mid-1940s, the black vote – concentrated in urban areas in states rich with electoral college votes – became increasingly crucial to Democratic success at the national level.

But beyond narrow reasons of electoral pragmatism were other factors that compelled the post–New Deal Democratic party to adopt the cause of black civil rights. The New Deal had firmly identified the Democratic party with the liberal intelligentsia in the northern states and with "social democratic" political currents. This did not sit easily with continued protection of the southern race system. The Second World War, fought against the explicitly racist regime of Adolf Hitler, made the defense of racial segregation and subordination in a section of the United States embarrassing in the international arena. Moreover, segregation of the military and several other incidents during the war years had exposed the harsher side of the southern social system to the national spotlight. In such circumstances, separate blood banks for black and white GIs, to take one example, were increasingly difficult to defend politically.[42] Finally, the advent of the cold war and America's emergence as the dominant world power exposed its domestic politics to worldwide attention. Again the sustenance of racial segregation in the South did not sit well with the promotion of "de-colonization" in foreign policy and efforts to woo the new "nonaligned" nations in the United Nations away from the Soviet orbit.

The inclusion of a weak civil rights plank in the 1948 Democratic platform was the first real challenge to southern segregation from the national Democratic party, and a gamble that the Democrats could consolidate their new electoral strength in the North while retaining enough of their hitherto-solid southern support to win in November. The gamble succeeded, for despite the defection of Alabama, Louisiana, Georgia, and South Carolina to South Carolina Governor Strom Thurmond's states' rights or "Dixiecrat" ticket, Harry Truman was still able to defeat Republican Thomas E. Dewey. V. O. Key's analysis of the presidential vote in 1948 reveals that support for the Dixiecrats closely paralleled support for Al Smith in 1928; that is, it was those portions of the South with the highest level of cotton production, the greatest numbers of black voters, and the most directly controlled by the traditional political elite that defected from the Democratic ticket to support Thurmond.[43] After 1948 the southern veto on presidential nominees who were prepared to challenge segregation was no longer operative (Roosevelt had persuaded the party to abandon the two-thirds rule in 1936), and the White South became increasingly uncomfortable with the Democratic party.

Social and economic changes inside the South were also gradually undermining Democratic presidential strength there, and it was those changes that laid the foundations for a viable southern Republican party during the 1950s. The South's economic and industrial development created a larger southern business class, which was clearly to the Republicans' advantage. Indeed, the growth in the Republican presidential vote in the South's new urban and economic centers—Atlanta, Houston, Dallas, Charlotte, and the Virginia suburbs of Washington, DC—was noticeable as early as the 1940s. These areas continued to grow apace in the decades that followed, and the urban middle-class vote in the South became the basis of Republican electoral strength.[44]

The GOP also benefited from the net effect of migrations to and from the South. Groups leaving the region in the postwar period tended to be rural, loyally Democratic, lower-status whites and blacks who could no longer make a living on the land, whereas those attracted to the South by the postwar business expansion tended to be urban, white, northern, white-collar, Republican-leaning business types.[45] This again weakened Democratic support in the urban and suburban areas of the South, particularly in the expanding "Rim South" states of Texas, Virginia, Florida, and North Carolina.

As it became apparent that the Democratic party nationally had become more committed than the GOP was to the cause of civil rights for blacks, the Republicans were able to add rural, lower-status whites to their upper middle-class, business base and to present a serious challenge to the Democrats in presidential elections in most southern states. Dwight D. Eisenhower carried Virginia, Texas, Florida, and Tennessee in 1952, and four years later he added Louisiana, becoming the first Republican presidential candidate to win a plurality of the southern popular vote. In 1960 Richard Nixon held Florida, Virginia, and Tennessee for the Republicans and was competitive in most of the region. Four years later Republican Barry Goldwater turned American post–Civil War electoral geography on its head by sweeping the five states of the Deep South—Mississippi, Alabama, Georgia, Louisiana, and South Carolina—and losing almost everywhere else. By 1964 it had become clear that white southerners, the traditional mainstay of the Democratic party, were increasingly prepared to vote for Republican candidates in presidential elections.[46]

Below the presidential level there were also signs of Republican growth in the south, albeit at a much slower pace. In the 1960s, Republicans began to contend seriously in statewide races and in elections to the U.S. Congress. In 1961 John Tower became the Republicans' first southern senator in the twentieth century when he won a special election for

Lyndon Johnson's Senate seat. During the 1950s and 1960s the proportion of Republicans elected in the South at all levels slowly began to rise, first in the Rim South and later, after 1964 and the final divorce between the segregationists and the Democratic party, in the Deep South as well (see Table 2.2).[47]

The final factor that needs to be discussed in this analysis of the breakdown of the Solid Democratic South is the role played by the black civil rights movement and their allies in the federal courts.[48] While these social and political trends were gradually moving the national Democratic party away from the white South, the final breakdown was greatly accelerated and precipitated by the political arousal and mobilization of blacks in response to federal court actions in their favor. The NAACP's strategy of using the federal judiciary to undermine segregation and black disfranchisement first began to pay dividends in 1944 when in the

Table 2.2. The Emergence of the Two-Party South, 1952 to 1968

	Percentage of Democrats				
	1952	1956	1960	1964	1968
Southern electoral college vote	55	47	63	63	20
Southern U.S. senators	100	100	100	91	82
Southern U.S. congressmen	94	93	93	85	75
Southern governors	100	100	100	100	96
Southern state senators	98	97	97	95	87
Southern state representatives	96	96	96	95	87

Sources: Compiled from data in Earl Black and Merle Black, *The Vital South: How Presidents Are Elected* (Cambridge, MA: Harvard University Press, 1992); and Jack Bass and Walter De Vries, *The Transformation of Southern Politics: Social Change and Political Consequence Since 1945* (New York: Meridian Books, 1977).

case of *Smith v. Allwright*, the U.S. Supreme Court declared the "white primary" to be unconstitutional.

The major breakthrough, however, was the *Brown v. Board of Education* decision of 1954, which outlawed segregation in education and overturned the *Plessey* precedent. In response to *Brown*, whites mobilized to resist implementation of the decision, and Martin Luther King, Jr., inspired black countermobilization to ensure that the decisions were enforced. Although the two major parties nationally still attempted to straddle the race issue during the 1950s, after 1960 the issues created by the civil turmoil in the southern states could no longer be ignored, and the national Democrats were forced to choose sides. In the 1960 campaign and during his administration, John F. Kennedy clearly identified the national Democratic party with King and the cause of black civil rights.[49] Kennedy's conflicts with the segregationist governor of Alabama, George Wallace, marked the final divorce between the Democratic party and the white South.[50]

In 1964 the segregationist South went for Goldwater instead of the Democratic candidate, since Lyndon Johnson had confirmed the Democrats' new position as the party of civil rights by his sponsorship of the 1964 Civil Rights Act, which finally began to dismantle the vestiges of segregation and racial discrimination in the South. Johnson's Voting Rights Act of the following year confirmed the political realignment in the South. All the devices intended to keep black voters from the polls were finally prohibited, and southern voting procedures were to be supervised by federal marshals. This, together with the outlawing of racial and antiurban gerrymandering of congressional and state legislative districts in the Supreme Court's *Baker* and *Reynolds* decisions, dealt the final blow to the impressive political edifice erected by the bourbon Democrats. With southern blacks now coming into the electorate as Democrats, one-partyism—already dying at the federal level—could no longer be maintained even in state and local politics. The stage was now set for the emergence of a two-party South, with political alignments based on class rather than race issues, as V. O. Key had envisaged in 1948.

In fact, although class issues have become more important in southern politics with the ending of segregation and disfranchisement, they have only occasionally predominated. Instead, class cleavages have been generally stifled by racial and cultural issues in southern politics, and the party system that has emerged has been characterized by a dual or a "split-level" political alignment. As the next chapter demonstrates, the South has become overwhelmingly Republican in presidential politics since 1964, primarily because of overwhelming support for Republican presidential candidates by white southern voters of all social classes. In congressional elections, however, the Democratic party's dominance persists (particularly in the House), and something like a biracial class coalition in favor of the Democrats has proved to be remarkably durable. In the next two chapters I explain this situation by looking at the relationship between southern Democrats and the national party at both the presidential and congressional levels, and I analyze the power of the southerners in the factional politics of the national Democratic party.

3

The Democrats' Presidential Weakness
in the South, 1968 to 1988

The South ceased to be solidly Democratic at the presidential level during the 1950s, and since 1964 it has been the Democrats' weakest region in presidential elections. Between 1968 and 1988 their consistently poor electoral performance in the southern states also doomed the Democrats' efforts to win nationally. The sole Democratic presidential triumph during that period (in the South and nationwide), in 1976, was the exception that proved the rule. It was impossible for the Democratic party to win presidential elections without winning at least some of the electoral votes of the southern states, yet during the 1980s this appeared to be an increasingly remote possibility.

Despite losing presidential elections in the South, the Democrats nevertheless remained the region's dominant party below the presidential level. Although the Republicans were competitive in senatorial and gubernatorial elections in the South by the late 1980s, they still held only about a third of the region's House seats, and the Democrats enjoyed overwhelming and secure majorities in almost all southern state legislatures.

This chapter explains one side of this paradox, the persistent weakness of the Democratic party in presidential politics in the South since 1964. The main emphasis falls on the party's revised nominating process, which for most of this period was structurally biased in favor of candidates from the party's neoliberal and New Left factions, with little appeal to most southern white voters and in which the South seemed unable to exercise any real leverage. Finally, I discuss various attempts by southern Democrats to remedy this situation.

The Years of Realignment, 1968 and 1972

In the 1968 and 1972 elections, the southern Republicans moved from a position of competitiveness in presidential elections in the southern states to one of complete dominance. The picture is somewhat complicated by the third-party presidential candidacy of the Alabama segregationist George Wallace in 1968, but by 1972 the new outlines of presidential politics in the South had clearly emerged.

The presence of George Wallace on the presidential ballot in 1968, with an appeal directed explicitly at the lower-status white constituency that had been attracted to Goldwater in 1964, somewhat restricted the Republican presidential advance. Wallace won four of the five Deep South states that had gone for Goldwater—Mississippi, Alabama, Georgia, and Louisiana—plus Arkansas. At the county level there was a high correlation between the support of Wallace in 1968 and that of Goldwater in 1964 and Thurmond in 1948, but Wallace also ran strongly outside the traditional black-belt segregationist bastions (where blacks were now voting in large numbers), in the upcountry areas, and among urban blue-collar southerners.[1]

Throughout the South, lower-status whites flocked to Wallace's banner, but his race-tinged populism had little appeal to the growing southern, suburban middle class that remained loyal to the Republicans and Richard Nixon.[2] Even in the three-way contest of 1968, Nixon was able to carry the Rim South states where he and Eisenhower had been strong a decade previously—Florida, Virginia, Tennessee, and North Carolina— and Nixon was also able to eke out a narrow victory in South Carolina with the help of GOP-convert Senator Strom Thurmond.

The remarkable feature of the 1968 presidential election in the South, however, was the collapse of the Democratic vote. Hubert Humphrey ran third in most southern states and was able to achieve a narrow plurality only in Texas, with heavy black and Hispanic support and the assistance of outgoing President Lyndon Johnson. White Southerners had departed the Democratic party en masse to vote for Nixon and Wallace.

After the 1968 election, Nixon aide Kevin P. Phillips wrote *The Emerging Republican Majority*, in which he predicted a Republican electoral realignment based on the election outcomes of 1968. Summing the Nixon and Wallace totals in 1968, Phillips predicted that the region would become predominantly Republican as the Wallace constituency of traditional southern whites eventually aligned themselves with the GOP as the only viable alternative to a Democratic party increasingly dominated by "new politics liberalism" on economic, racial, cultural, and foreign policy issues. The Democratic party was no longer a fold within which southern conservative whites could find a place, and the Democrats' southern support would be confined to the black and Hispanic minorities and a smattering of white liberal enclaves such as the southern college towns and southern Florida.(see Table 3.1)[3]

Events in the 1972 presidential election worked out almost exactly as Phillips had predicted. Wallace returned to the Democratic camp and ran strongly in the early primary elections both inside and outside the South, but the attempt on his life in May 1972 wounded him seriously enough to eliminate him as a serious presidential prospect in that year. Reforms in the Democrats' presidential nominating process and the continuing liberal agitation against the Vietnam War delivered the presidential nomination to the candidate of the "new politics" (representing the temporary alliance between the neoliberals and the New Left), South

Table 3.1. The Republican Presidential Realignment in the South, 1968 to 1972 (percent)

State	Nixon 1968	Wallace 1968	Humphrey 1968	Nixon + Wallace	Nixon 1972
Rim South:					
Virginia	43	24	33	67	69
Florida	41	29	31	70	72
Texas	40	19	41	59	67
North Carolina	40	31	29	71	71
Tennessee	38	34	28	72	70
Arkansas	31	39	30	70	69
Deep South:					
South Carolina	38	32	30	70	72
Georgia	30	43	27	73	75
Louisiana	24	48	28	72	70
Alabama	14	66	19	80	74
Mississippi	14	64	23	78	80

Sources: Adapted from data in Kevin P. Phillips, *The Emerging Republican Majority* (Garden City, NY: Anchor Books, 1970) p. 208; and Jack Bass and Walter De Vries, *The Transformation of Southern Politics: Social Change and Political Consequence Since 1945* (New York: Meridian Books, 1977), p. 403.

Dakota Senator George McGovern, whose electoral appeal to the South was negligible outside the solidly Democratic black community.[4]

In the fall campaign Nixon made an explicit appeal for the Wallace vote and portrayed his Democratic opponent as the candidate of "amnesty, acid, and abortion." On election day Phillips's prediction came true as the Republican presidential candidate swept all the southern states by lopsided margins, and apart from the 1976 aberration, the South remained solidly in the Republican column until 1992 (see Table 3.2).

Analysis of a Realignment

The declining identification with the Democratic party among white southerners was the key to the southern presidential realignment.[5] And although the Democrats were partly compensated by the overwhelming support of southern blacks after they were admitted to the electorate in significant numbers after the 1965 Voting Rights Act, this was insufficient to make up for the movement of the overwhelming majority of southern whites to the GOP.[6] Even in 1976, Gerald Ford still carried a majority of the southern white vote, although when added to his formidable performance among blacks, Jimmy Carter got enough to carry every southern state but Virginia.[7] This dire performance among white southerners became the Democratic party's gravest problem in presidential politics, since without winning more of these voters, the Democrats could not win the southern states they required in order to envisage presidential success on a national scale. According to Earl and Merle Black:

Table 3.2. Presidential Republicanism in the South, 1964 to 1988

State	Number of Elections, 1964–88	Number of Republican Victories	Percent
Alabama	7	5	71
Arkansas	7	4	57
Florida	7	5	71
Georgia	7	4	57
Louisiana	7	5	71
Mississippi	7	5	71
North Carolina	7	5	71
South Carolina	7	6	86
Tennessee	7	5	71
Texas	7	5	71
Virginia	7	6	86

The conversion of the South into a Republican stronghold represents a fundamental change in the structure of presidential politics . . . For all presidential elections through 2000, sweeping the South would permit the Republicans to win the presidency with slightly more than three-tenths of the northern electoral vote. Conversely, the loss of southern electoral votes means that the Democratic party now must win a higher percentage of the North's electoral vote (69 percent if no electoral votes are won in the South) than the Republicans once needed when they had to write off the South.[8]

In place of their old lily-white coalition, the post–Voting Rights Act Democrats needed to construct a biracial "class" coalition of blacks and lower-status whites to defeat the white, middle-class GOP. Although they succeeded in accomplishing this only once in presidential politics between 1968 and 1988, it was, nevertheless, the persistence of such a biracial coalition that accounted for their remarkably consistent success in southern elections below the presidential level (see Table 3.3).[9]

There was thus no overall Republican realignment in the South after the civil rights revolution, since the GOP conspicuously failed to replicate its dominance at the presidential level at other levels of government. Indeed, in the state legislatures, Democratic dominance—with one or two exceptions—remained overwhelming. The South evidenced the split-level realignment that characterized American national politics after

Table 3.3. Democratic Strength at Different Electoral Levels in the South in 1988 (percent)

State	Democrats for President	Democrats for U.S. Senate[a]	Democrats for Legislature[b]
Alabama	40	–	85
Arkansas	43	64[1]	88
Florida	39	50	60
Georgia	40	–	80
Louisiana	45	–	84
Mississippi	40	47	91
North Carolina	42	44[1]	65
South Carolina	38	–	72
Tennessee	42	66	61
Texas	44	60	64
Virginia	40	71	68

[a]For Arkansas and North Carolina the figure is the Democratic percentage in the governor's race.

[b]The Democratic percentage of the total membership of the state legislature after the 1988 elections.

Source: National Journal, 12 November 1988, 2847, 2866, 2884, 2886.

1968, but in an exaggerated form: The Republican presidential dominance was greater, and the Democrats retained an apparently impregnable advantage at the lower electoral levels.

This curious state of affairs was largely due to the electoral behavior of the swing group in southern electoral politics: lower-status whites. The latter habitually abandoned the Democrats in presidential politics, but in lower-level contests, enough of them "came home" to enable a biracial Democratic coalition to prevail. This coalition was the key to the successful election of the "New South" generation of governors during the 1970s, as their populist "new politics" appeal often succeeded in winning over alienated poorer whites who had inclined toward Goldwater and Wallace during the 1960s.[10] Just as Carter triumphed in the 1976 presidential race with such a strategy, so many southern Democratic politicians of a similar stamp were able to triumph in state politics. As Alex Lamis wrote:

> Thus as the race issue was fading, skillful Democratic politicians who were often successful products of their state's political institutions lost no opportunity to appeal to the great "reservoir of Democratic party allegiance" that [Jimmy] Carter had spoken of and that the survey data measured. . . . Although these politicians took various stances on the issues, rarely did their positions form a coherent approach that allowed them to be easily labeled in recognizable partisan terms. And through this straddling process, they managed to remain acceptable to white traditional Democrats, who preferred to stay with their party if it was at all possible. These astute Democratic politicians with their innocuous campaigns made it possible.[11]

The swing constituency of poorer whites remained highly responsive to populistic Democratic appeals on economic issues, despite their distaste for the Democrats' post-1968 foreign policy stances and social liberalism.[12] Moreover, the Democratic party's greatest strength since the New Deal has been its ability to deliver services to constituents using the advantages of incumbency at both the federal and state levels. Economy-minded Republicans are less likely to be so protective of services and programs, and thus contemporary southern Democrats still generally speak to the economic concerns of lower-status white voters much more convincingly than most Republicans can. Finally, southern Democratic candidates for congressional, state, or local office are more likely to have strong local roots and, therefore, a political style that is more familiar and attractive to traditional white voters than is that of a suburban Republican born outside the South.

The situation of the lower-status southern whites after the civil rights movement is often described as one of "alienation" and "dealignment."[13] Undoubtedly this section of the electorate was the most cross pressured in the South: pulled toward the Democrats on economic issues and toward the GOP on defense and social questions. During the 1970s and 1980s these voters resolved their dilemma quite rationally by consistently splitting their votes. In presidential elections, in which defending the nation and setting an appropriate national tone appeared to be more significant, they tended to vote Republican, whereas in congressional and state races that dealt mainly with services and programs with a direct impact on their personal economic situation, they continued to favor the Democrats. By thus avoiding ideological consistency, they got the best of both worlds.[14]

Of course, this presented real problems to the Democrats in trying to win presidential elections in the South. However, as most of the lower-status whites had not firmly realigned themselves as Republicans, there was ostensibly no reason that if the Democrats produced presidential candidates similar to those they produced in congressional and state elections in the South, they could not also do better in presidential elections there.

This brings us to the second set of factors in the Democrats' presidential weakness in the region, the inability of their nominating process to generate candidates with real appeal to white southerners. Unfortunately for southern Democrats, the party's nominating process after 1968 operated against such a candidate's emerging as the party's nominee and markedly in favor of candidates who had the least appeal to traditional white southern voters.

The South and the Democrats' Presidential Nominating Process Since 1968

The first factor that undermined southern influence over the Democratic party's presidential nominating process was the abolition of the two-thirds rule in 1936. This was done to prevent another deadlocked marathon that reflected badly on the entire party as in 1924, but the effect on the South's ability to veto candidates of whom it disapproved was profound, although the implications did not become fully clear until 1948. The surprising thing was that the southerners did not apparently perceive the threat to their power involved in the abolition of the rule, and so it expired with surprisingly little resistance at the 1936 convention.[15]

The process of reforming the nominating process over the next half-century continued to undermine the power of the white South within the party. Abolition of the whites-only primary after the Supreme Court's 1944 *Smith v. Allwright* decision and the gradual enfranchisement of blacks in delegate selection processes were direct assaults on the old white power structure. In 1964 the presence of lily-white national convention delegations from the South at the Democratic National Convention had become a grave embarrassment to the party, and the struggle for black representation came to a climax at the 1968 convention with the momentous decision to seat the integrated Mississippi Freedom Democratic delegation in place of the lily-white regular delegation.[16] National rules were now beginning to undermine the solidity of the southern delegations.

These changes were confirmed by the further reforms of the nominating process introduced by the McGovern–Fraser Commission between the 1968 and 1972 conventions. This commission had been established explicitly to appease the Democrats' "new politics" liberals who had been outraged at the failure of the anti–Vietnam War forces, led by Senator Eugene McCarthy, to prevail at the 1968 Democratic convention. Having been thwarted by an alliance of the traditional Democratic factions—southern conservatives and northern regulars—the New Left and the neoliberals used the party's reform commission to restructure the presidential nominating process in a manner favorable to their own interests. The most surprising feature of the McGovern–Fraser Commission's wholesale assault on the national Democratic power structure was that it occasioned so little outright resistance from the party's other major factions, which had a great deal to lose.[17]

The McGovern–Fraser proposals forced the state parties to make their delegate selection procedures as "open" and "representative" as possible and also required "women, minorities, and youth" to be represented in a state's convention delegation in accordance with their presence in the state's general population. Finally, the commission prohibited the use of the unit rule.[18]

Rather than risk handing over the entire state party apparatus to proponents of the "new politics," most state party leaders, both inside and outside the South, preferred to respond to the commission's guidelines by establishing a presidential primary election.[19] Between the 1968

and 1988 the number of presidential primaries inside the South increased
considerably (see Table 3.4).

The net effect of all these changes was to shatter southern solidity for
good, by fragmenting the South in Democratic presidential nominating
politics. Southern convention delegations were now divided between
blacks and whites, liberal and conservatives, and men and women and
among the devotees of different presidential candidates (in large part as a
result of the McGovern–Fraser Commission's requirements of proportional representation).[20] Although blacks and liberal white southerners
gained as a result of the changes, representatives of the more traditional
white southerners lost out. Indeed, the South almost ceased to be a
distinctive factor in the Democratic presidential nominating process.[21]

Southern candidates and candidates with potential appeal to the white
South also suffered from the structure of the new primary-based nominating process. To mount a serious challenge for the nomination, it
became necessary for presidential aspirants to perform creditably in the
crucial early contests in Iowa and New Hampshire.[22] Democrats in these
states, however, did not tend to favor the kinds of Democratic candidates
that appealed to the white South. Iowa's political culture is isolationist,
pacifistic, and progressive, and New Hampshire Democrats have tended
to favor neoliberal candidates—Eugene McCarthy, George McGovern,

Table 3.4. Processes for Selecting the Democratic National Convention
Delegation in the Southern States, 1968 and 1988

State	Process Used in 1968	Process Used in 1988
Alabama	Delegate primary[a]	Presidential primary
Arkansas	State party committee	Presidential primary
Florida	Presidential primary	Presidential primary
Georgia	State party committee	Presidential primary
Louisiana	State party committee	Presidential primary
Mississippi	Caucuses	Presidential primary
North Carolina	Caucuses	Presidential primary
South Carolina	Caucuses	Caucuses
Tennessee	Caucuses	Presidential primary
Texas	Caucuses	Presidential primary
Virginia	Caucuses	Presidential primary

[a]Primary to elect delegates only, with no presidential preference poll.

Sources: Austin Ranney, *The American Elections of 1980* (Washington, DC: American Enterprise
Institute, 1981), app. D, pp. 366–68; and Rhodes Cook, "The Nominating Process," in Michael Nelson,
ed., *The Elections of 1988* (Washington, DC: Congressional Quarterly Press, 1989), p. 30.

Gary Hart—as opposed to those associated with the traditional Democratic coalition of ethnics, labor, and the white South. In short, candidates from the party's neoliberal faction had a "built-in" advantage in the revised presidential selection process.

Because of the effects of media "momentum" on the candidates' visibility and fund-raising, candidacies that fail to register at least a top-two finish in Iowa and, particularly, New Hampshire tend to be still-born.[23] By the time that the primary season reached the South, the also-rans in New Hampshire were invariably no longer regarded as serious contenders. The southern or southern-oriented candidacies of John Glenn, Fritz Hollings, and Reuben Askew all died in New Hampshire in 1984. In 1988 a similar fate befell Richard Gephardt, although he had won in Iowa a few weeks previously. As we shall see, the South concentrated its 1988 primaries on one day in order to try to distract the media's attention from Iowa and New Hampshire and to increase its regional clout. Moreover, one southern champion in 1988, Tennessee Senator Albert Gore, pointedly ignored Iowa and New Hampshire in order to concentrate on gaining a delegate windfall in the southern states.

It was thus evident that the early primaries in unfavorable terrain for southern contenders usually acted to create Democratic front-runners clearly identified with the northern, neoliberal section of the party and with very limited appeal in the South, a factor that militated heavily against the party's prospects in the region in November.

Jimmy Carter, the Exceptional Case

Up to this point I have deliberately ignored the exceptional case that contradicts everything discussed so far. That is, in 1976 Jimmy Carter won in Iowa and New Hampshire, swept the South in the primaries, was nominated, and won the presidency in November largely owing to an almost solidly Democratic South.

Jimmy Carter's long-shot campaign for the presidency in 1976 succeeded because of good planning, good fortune, and the context of the election. Carter's managers appreciated the implications of the move to a primary-centered nominating campaign better than did those of his opponents, and Carter was off and running early and everywhere after leaving the Georgia statehouse in 1974. The Carter campaign further understood the critical nature of the Iowa caucuses and the New Hampshire primary and invested a great deal of time in both states.[24] Carter also had the moderate-to-conservative side of the Democratic party all to himself in those early contests while his opponents divided-up the regular/neoliberal/New Left vote. Having established himself early as the front-runner after New Hampshire, Carter united blacks, liberal whites, and a sufficient number of lower-status white Democrats attracted by his populist, "antipolitics" themes to defeat George Wallace in the Florida primary. From that point on his nomination was virtually assured.

Yet even given the Carter campaign's excellent strategic sense, he could not have prevailed but for the unique circumstances of the 1976 election. For a country traumatized by the effects of Vietnam and Watergate, Carter's "government of love" had a particularly significant resonance. By capitalizing on the general revulsion toward politicians, Carter was also able to avoid taking the more explicit positions on issues that might have doomed his campaign in the early primaries. To conservative Democrats he sounded like a conservative, and even though neoliberals and New Leftists might have preferred Morris Udall to Carter, they far preferred Carter to the segregationist George Wallace or to veteran regulars like Scoop Jackson and Hubert Humphrey.

Another unique attribute of the Carter campaign was Carter's strong appeal to black voters in both the North and the South. Carter's Southern Baptist background ironically was probably his greatest asset in cultivating black voters. The presence of key black political spokespersons from Georgia such as Rev. Andrew Young, Coretta Scott King, and Rev. Martin Luther King, Sr., in the Carter entourage also enhanced Carter's appeal to black Americans.[25]

In the general election, Carter became the first Democratic presidential contender in recent years to obtain a respectable southern white vote, sufficiently respectable indeed when added to his formidable black voting base, to carry him to victory in November over Gerald Ford in every southern state save Virginia, although Ford still secured a majority of the southern white vote.[26]

Carter's coalition in both the nominating and the general election campaigns demonstrated a clear route by which a white Democrat from the Deep South could win the presidency. Yet it was a coalition that probably only the post-Vietnam and Watergate mood of 1976 could have produced, and it was fundamentally unstable. For instance, Carter won despite his southern background not because of it, and he certainly did not run in the early primaries as Dixie's candidate—had he done so, he would probably not have fared as well as he did in Iowa and New Hampshire. It was also hard to envisage another white southern candidate with as powerful appeal as Carter had to both blacks and lower-status whites in the South, particularly as black candidates are likely to be a regular feature in future Democratic presidential contests. Lower-status whites were attracted to Carter because they saw him as a more physically vigorous version of their ailing hero George Wallace, and it should be noted that after defeating Wallace in Florida, Carter took pains to give no offense to the Alabama governor or his supporters during the remainder of the 1976 nominating campaign.[27] By straddling or evading the policy issues that divided the races in the South, Carter was able to put together a coalition of blacks and lower-status whites alienated from Washington politics since the mid-1950s.

Once in office, President Carter could no longer avoid taking clear positions on these issues, as he had done in the 1976 campaign, and given the nature of his administration and that of the national Democratic party, it was inevitable that he would fall on the side of social liberalism on most questions while neglecting the economic populism that had won him the South in 1976. But by doing so, he alienated the southern white component of his coalition sufficiently to ensure his defeat in the region and in the nation in 1980. Whatever their merits, the Panama Canal Treaty, the energy policy, the emphasis on promoting the rights of women and minorities—achievements in which Carter took special pride—were not particularly well received in the white South. By the late 1970s, evangelical white southerners who had supported Carter as one of their own in 1976 had largely deserted the administration and the Democratic party.[28]

Ronald Reagan, by emphasizing issues such as the "sellout" of the Panama Canal, abortion, school prayer, busing, and the restoration of national pride, evoked a powerful response in the white South. In the 1980 election he restored the post-1968 Republican presidential advantage in the region (although his victory margins in several of the Deep South states against the regional favorite were narrow), and the Republican grip was reinforced in 1984.[29] Carter was an aberration in the pattern of Democratic nominees, and 1976 was an aberrant election. In office, when the white South saw Carter as just another liberal Democrat—albeit one with a southern accent—the brief Democratic revival in southern presidential politics was quickly extinguished.

The Super Tuesday Debacle in 1988

In 1984 the compression of the primary election season demanded by the
Democratic National Committee led to the states of Florida, Georgia,
and Alabama holding their primaries on the same day (11 March).
Theoretically this should have been an advantage to the three more
conservative contenders for the presidential nomination – Senator John
Glenn of Ohio, Senator Ernest Hollings of South Carolina, and former
Governor Reuben Askew of Florida – who might reasonably have hoped
for a windfall of delegates and favorable publicity from a series of
triumphs in the southern region. In reality, however, the first "Super
Tuesday" was a victim of the media momentum generated by the Iowa
caucuses and the New Hampshire primary. Those two states had produced a tight contest between Walter Mondale – representing the regular
wing of the Democratic party – and Gary Hart – representing the neo-
liberalism of McCarthy, McGovern, and, to some extent, Carter. Neither
was very attractive to the white South, but after the New Hampshire
primary, they were generally treated by the news media as the only two
viable contenders. Indeed, the Hollings and Askew campaigns were
terminated before Super Tuesday! The short space of time between New
Hampshire and the three southern states' primaries, together with the fact
that some important northern states (principally Massachusetts) were
holding their primary on the same day, led to a whirlwind campaign, with
Hart, Mondale, Glenn, and Rev. Jesse Jackson (representing the New
Left/minorities section of the party) hopping madly from airport to
airport to create "sound bites" for local news in the various southern
media markets.[30]

After Mondale was routed by Reagan in the November election,
southern state party leaders and Democratic elected officials gave serious
thought to how the South could increase its leverage over the nominating
process and in the national councils of the party. They worried that
continuing electoral obliteration at the presidential level would eventually filter down to the lower electoral levels and that many conservative
white Democratic officeholders might defect to the GOP in order to
salvage their political careers.[31] These concerns led a group of prominent
southern and border state politicians, such as Senator Sam Nunn, Senator
Charles Robb, and Congressman Richard Gephardt, to form the Democratic Leadership Council (DLC), in an effort to steer the Democrats'
policy and image away from close identification with liberal interest

groups (feminists, minorities, labor, teachers, gays) and "weakness" on defense issues.[32] A 1985 special election, in which a Democrat held off a strong Republican challenge in a rural, white, Texas congressional district by using the issue of protectionism, seemed to indicate that by combining a more resolute image on foreign and defense policy with economic populism, the Democrats might be able to break the Republican hold on the white South and to reconstruct their biracial coalition on a national level.[33]

Optimism about Democratic prospects in the South increased after the 1986 midterm elections, when the Democrats regained Senate seats in Georgia, Alabama, North Carolina, and Florida that they had lost in 1980 and held on to what initially seemed to be a very vulnerable seat in Louisiana. Campaigning on economic populist themes, even Democrats with a liberal political profile, such as the new Georgia senator, Wyche Fowler, were able to build black–white coalitions and get elected.[34] (Of course, it might be added that the crop of defeated incumbent Republican senators elected in the 1980 Reagan sweep was not particularly impressive.)

As the centerpiece of their strategy to increase southern leverage over the 1988 Democratic presidential nominating process, the DLC and southern state and local party leaders devised the idea of a truly "Super" Tuesday, concentrating all of the South's primaries and caucuses on one date in March. By rigging the primary schedule in such a fashion, the devisers of Super Tuesday reckoned that they would finally dilute the influence of Iowa and New Hampshire and provide an opportunity for a candidate more representative of southern opinion to emerge at the end of the process.[35] On paper the idea seemed very good. But in practice the authors of Super Tuesday might have learned from the more limited concentration of contests in 1984 that the outcome of their scheme might be quite different from what they had intended (see Table 3.5).[36]

That is, the southern leaders had failed to appreciate that the more primaries that were held on just one date just after Iowa and New Hampshire, the more important that Iowa and New Hampshire would become, since these two small northern states still determined which candidacies would be judged viable heading into the southern regional primary. Thus in 1988 the clear Democratic front-runner heading into Super Tuesday with the full benefit of media momentum was the New Hampshire victor, Massachusetts Governor Michael Dukakis, a northeastern representative of the Democrats' neoliberal faction and hardly the

Table 3.5. The Overall Result of the Super Tuesday Primaries in 1988

Candidate	Number of Primaries Wonª	Percent of Total Vote	Number of Delegates Won	Percent of Delegates
Dukakis	5	26	356	32
Jackson	5	27	353	32
Gore	5	26	318	28
Gephardt	1	13	94	8
Others	0	9	0	0
Total	16	100	1121	100

ªOn 8 March 1988, primaries were held in ten of the eleven southern states (South Carolina held a caucus on another date) plus Kentucky, Maryland, Massachusetts, Missouri, Oklahoma, and Rhode Island. The figures in the table do not include the returns from the states that held caucuses on Super Tuesday.

Source: Adapted from tables in *Congressional Quarterly Weekly Report*, 12 March 1988, pp. 636–38.

kind of southern favorite the authors of Super Tuesday had envisaged.[37] The most explicitly southern and conservative contender, Senator Albert Gore, avoided Iowa and New Hampshire entirely so as to concentrate on Super Tuesday, but by doing so he inevitably suffered in terms of momentum and viability, which in turn affected his Super Tuesday performance.[38] Another candidate with potential appeal to the South, Congressman Richard Gephardt, succeeded in narrowly winning the Iowa caucuses, but he had to devote so much of his money to campaigning in Iowa and New Hampshire that he did not have sufficient funds remaining to mount the necessary TV-advertising campaign in the southern media markets required for Super Tuesday. In a different way he was another victim of the "Big Mo" or the search for it (see Table 3.6).[39]

Another factor from 1984 that recurred in 1988 was several significant northern states' (particularly Massachusetts and Maryland) scheduling their primaries on the same day as Super Tuesday (8 March). This reduced somewhat the attention that candidates could give to the South and also gave an additional boost to Dukakis, the clear front-runner in both states, since he could claim some substantial victories even if he won nothing in the South and thus maintain his momentum. Moreover, the holding of so many primaries and caucuses on the same day served only to dilute the attention received by individual southern states.

The next difficulty was that the two most obviously southern contenders, Gore and Gephardt, were perhaps not the most effective champions that Dixie could have put forward. Both had fairly liberal voting records in Congress, and neither really aroused a great deal of enthusiasm among

Table 3.6. The 1988 Democratic Primary Vote by Time Periods (percent)

Candidate	Before Super Tuesday	Super Tuesday	After Super Tuesday	Final Outcome
Dukakis	39	26	54	42
Jackson	11	27	31	29
Gore	6	26	5	14
Gephardt	24	13	1	6

Source: Adapted from figures in Rhodes Cook, "The Nominating Process," in Michael Nelson, ed., *The Elections of 1988* (Washington, DC: Congressional Quarterly Press, 1989), p. 53.

activists in the region. Gephardt had some appeal to lower-status white southerners with his populistic protectionism, and Gore tried to depict himself as the only real defense "hawk" in the Democratic field (not a particularly difficult task in 1988, it should be added) but neither really struck a profound cord in the South in the way that Wallace, Carter, or Ronald Reagan had done. Despite having helped in the creation of the DLC and Super Tuesday, more authentic southern Democrats such as Governor Charles Robb of Virginia or Senator Sam Nunn of Georgia refused to enter the race, showing a certain lack of faith in their own creation.[40] The final problem with the Gore and Gephardt candidacies was that they tended to undercut each other, as both were aimed at the more conservative white vote. In the absence of one of them, the other might have made much more out of the opportunity of Super Tuesday than either of them did individually.[41]

Another factor that the framers of the southern regional primary appeared to have overlooked was that the southern Democratic primary electorate was no longer so socially and ethnically homogeneous as to be readily "delivered" for a regional champion of a moderate-to-conservative stamp. Conservative white southern voters largely stayed away from the polls on Super Tuesday (or voted in Republican presidential primaries where they had the chance), and exit poll data showed that the Democratic primary electorate in 1988 was disproportionately "liberal" in orientation, compared with the general electorate in the southern states.[42] In these circumstances, the Super Tuesday victories of Jackson and Dukakis appear less surprising.

In 1988 the candidate of the Democrats' New Left, Jesse Jackson, did even better among southern blacks than he did in 1984, so well, in fact, that thanks to the splintering of the white vote among various candidates, Jackson was able to win five Deep South states and a harvest of delegates (see Tables 3.5 and 3.6).[43] Indeed, Jackson was the real winner of Super Tuesday, since these successes enhanced greatly the viability of his candidacy in subsequent northern primaries.

Aside from the black vote, there were also substantial numbers of liberal Democratic enclaves in the South – particularly in the large metropolitan areas and academic communities of Florida, Texas, Georgia, and North Carolina – that responded very warmly to the Dukakis candidacy.[44] With his southern opponents undercutting one another and Jackson getting virtually all of the black vote, Dukakis, the northeastern liberal, was able to carry the two largest southern states, Florida and Texas (in the latter he also got a lot of support from Mexican-American voters). This further increased his momentum and created a false impression of national strength. Gore was able to carry Arkansas, Kentucky, North Carolina, Tennessee, and Oklahoma, but these successes were obscured by the strength of Jackson and Dukakis in other parts of the South. Gephardt won only his home state of Missouri, although it should be reemphasized that had he and Gore not been competing for the same voters, one of them might have achieved a much more convincing southern sweep.

The other question that arose in 1988 was whether a southern favorite could win the nomination even if he did emerge with a solid bloc of delegates and front-runner status after Super Tuesday. To win in the South might have made a candidate become too closely identified with that region for the comfort of more liberal Democratic primary voters in the North and West. A southern candidate might also adopt positions on issues so incongruent with those of Democratic activists nationally that they would be unable to convert Super Tuesday success into a successful nomination. The experience of Albert Gore in 1988 is instructive in this regard. Gore had positioned himself so far to the right (relatively speaking) on defense issues and identified himself so closely with the South in his effort to gain some momentum from Super Tuesday that he alienated potential support in the midwestern and northern states' primaries that immediately followed.[45]

The final problem with a regional primary to enhance southern leverage over the Democratic nominating process was that if the ploy did succeed, other regions might quickly follow suit with their own regional primary to increase their own influence in the process vis-á-vis the South. (There were hints of this in 1988, when the New England states and some midwestern states also considered coordinating the dates of their primaries). From four or five regional primaries it might be but a short distance to a national primary (as the logical culmination of rationalizing the presidential nominating process). After one successful Super Tuesday, there might not be too many more, and then where would the South be?

Conclusion: Waiting for Recession?

The Democrats' presidential problem in the South during the 1980s appeared unlikely to be easily resolved in the short term. The southerners knew what they had to do: build a black–white coalition around economic issues, as they had done at the congressional, state, and local levels in the South. But the problem they faced was that the major issues of presidential politics were far more divisive for that coalition than were state issues or the service issues that predominated in congressional elections. On the major issues of presidential politics during the 1970s and 1980s, white southerners tended to be interventionist in foreign policy and more conservative on moral and law-and-order questions (which a president can influence through the Justice Department and federal court appointments), and they identified strongly with national symbols closely associated with the presidency (such as the flag). Outside black areas and liberal enclaves, southern voters in presidential elections felt very uncomfortable with the Democratic presidential candidates, including one of their own, Jimmy Carter, when they discovered that he was not quite what they had expected.

To a large extent, the structure of the Democratic nominating process also discriminated against the South's interests. In the primary system the Democratic neoliberals, single-issue and advocacy groups, and the news media had much more influence than they did in the era of brokered conventions and the unit rule. In the old system of brokerage, the South exerted considerably more leverage over the national party in candidate selection, whereas in the post-1968 system, the South found it much more difficult to exert influence, primarily because in the new politics the outcome was determined more by interest groups, candidate organizations, and media expectations than by mobilizing local, state, or regional party leaders. Super Tuesday was an attempt to reassert southern leverage in the reformed nominating system, but its failure in 1988 illustrated why the whole logic of that system and the logic of regarding the South as a

separate entity with particular interests were in conflict. The new nominating process operated according to national criteria through national
media and tended to suppress regional effects, even in the most culturally
distinct region of the country. In this sense, southern politics and society
had become similar to that of the rest of the United States, and it was too
late to go back to the old Solid South model.

On the other hand, the Democrats certainly needed to win some
southern states to have any chance of reaching the White House. To
achieve this end they also needed a downturn in the economy (as in
1974–75) to attract lower-status white voters on bread-and-butter economic issues. Economic populism still had considerable electoral potential in the South, particularly when much of the region remained largely
unaffected by the southern economic boom.

In congressional and state elections that focused on how effectively
incumbents had served their states or districts, Democrats retained a
clear advantage throughout the South over the more parsimonious Republicans. Given a national recession, the prospects of a renewal of the
presidential Democratic party in the South were therefore quite good. In
such circumstances the themes of the 1988 Gephardt primary campaign –
economic populism with a nationalist/protectionist flavor that neutral-
ized Republicans' attacks on Democrats' patriotism – might well prove to
be very electorally effective in the southern states.

In the recession election of 1992, Governor Bill Clinton of Arkansas
confirmed this hypothesis by using precisely such populist/nationalist
themes to convert Super Tuesday into a successful campaign for the
presidential nomination by a southern Democrat (although not a particularly conservative one). The reasons for Clinton's success where other
southerners had previously failed in 1984 and 1988, together with his
breaking of the Republican stranglehold on the South in November, are
discussed in Chapter 6.

4

Southern Democrats in the U.S. Congress

Scholars of the U.S. Congress over the past two decades are virtually unanimous in their agreement that there has been a transformation in the internal power structure of both the House and Senate.[1] One aspect of this transformation has been a reduction in the influence of the southern Democrats. Under the congressional power structure that prevailed from 1910 to 1970, southern Democratic members exercised a dominant influence on the behavior and policy output of the Congress. Over the past two decades, however, this influence has clearly eroded.

This chapter deals with the situation of southern Democrats in the present-day Congress and is based largely on a series of interviews conducted during the summer of 1990 with thirty southern Democratic congressmen, three senators, two former senators, and five senior Senate staffers. The first section deals with the House and the second with the Senate.

In the each section I briefly describe the nature and sources of southern influence over the House and Senate between 1910 and 1970 and why the old congressional power structure broke down in the early 1970s. I discuss the reasons for the southern Democratic members' continued adherence to the Democratic party, and I then analyze how and why the position of the southern Democrats has evolved on various issues, particularly civil rights. I also examine the influence of the southern Democrats over the House and Senate power structure today and the effectiveness of several attempts to organize the southern Democrats in Congress. Each section concludes with a discussion of the tense relationship between the southern Democratic congressmen and senators and the national party, and the future prospects for the southern Democrats in each house.

The House

The Rise and Fall of the "Old South" in the House

The southern Democrats' grip on the House's power structure between 1930 and 1970 was due to four main factors: the power of the committee chairmen, the seniority rule, the dominance of the Democratic party in House elections since 1932, and the fact that the safest Democratic House districts were in the southern states.[2] These factors all interacted to produce a pronounced southern influence over the entire chamber. This influence was, of course, particularly salient in relation to civil rights issues. Defense of segregation provided the rationale for the establishment of the one-party Solid South and helped maintain it.[3] And it was this unrelenting opposition to civil rights bills that identified both the South and the Congress as bastions of conservative reaction.

This was only part of the story, however. Beyond the defense of segregation, there were significant divisions among the southern members on a variety of issues, particularly government intervention on the economy, national security, and welfare. On foreign and defense policy the southerners tended to be more internationalist during the 1930s and 1940s than were members from other regions, and on the economic issues of the New Deal, many—such as Alabama's Lister Hill and Texas's Maury Maverick and Lyndon Johnson—were staunch supporters of Roosevelt.[4]

Yet even given that the southern Democrats were not as monolithically conservative on other questions as they were on civil rights, on the economic and social welfare issues of the time a majority of the southern members—reflecting the generally rural and small-town nature of their districts—tended more toward the Republican position. This provided the basis for the so-called conservative coalition of Republicans and southern Democrats that was originally constructed around opposition to Roosevelt's "court-packing" bill in 1937.[5] The alliance between the economic conservatives of both parties generally prevailed on most fiscal and social welfare questions from the late 1930s up to the period of President Lyndon Johnson's "Great Society" legislation in 1965/66, when it could finally be outvoted by the northern liberal Democrats.[6]

The southern Democratic domination of the House came to an end because the southern social and political system that had sustained it collapsed during the 1960s. The civil rights revolution and the social effects of rapid economic development led to the emergence of two-party politics in the South, and the most conservative elements of the Democratic party gradually moved toward the nascent southern GOP. As a consequence the southern and border-state proportion of the House Democratic caucus shrank markedly, from 54 percent in 1948 to under 40 percent by 1980, while the percentages of House Democrats from other regions correspondingly increased (see Table 4.1). The impact of reapportionment after the Supreme Court's *Baker v. Carr* decision in 1962 also reduced the number of conservative rural districts that had sustained the careers of traditional southern House Democrats and increased the number of more economically and socially liberal urban and suburban districts.[7]

This accounts for the reduction in southern conservative influence in the House Democratic caucus, but it was the deliberate weakening of the committee seniority system and the transformation of the House's power structure that dealt the final blow. Younger Democrats from both North and South felt that the decline of party loyalties in the electorate required them to carve out more of a personal base of support in their districts, which, in turn, required that they gain more control over the congressional agenda for themselves. The authoritarian and aged southerners who controlled the key committees of the House stood as obstacles to their goals, and the younger, more liberal Democratic members now had

Table 4.1. Democratic House Seats by Region, 1924 to 1988 (percent)

Region	1924	1936	1948	1960	1972	1980	1988
South and Border	70	42	54	50	41	39	39
Plains and Midwest	11	24	16	15	17	22	23
New England and Middle Atlantic	15	23	21	22	25	25	21
Mountain and Pacific	4	11	10	12	17	16	18
	100	100	100	100	100	100	100

Source: Norman J. Ornstein, Thomas E. Mann, and Michael J. Malbin, *Vital Statistics on Congress, 1989–1990* (Washington, DC: American Enterprise Institute, 1990), pp. 11–12.

the numbers to effect change by using the hitherto almost defunct House Democratic caucus.

The effect of various reforms passed by the House Democratic caucus during the 1970s was to end the domination of major House committees by their chairmen, by devolving much of their power to subcommittee chairs and individual members. The seniority rule was considerably diluted by making all committee chairs potentially subject to a confirmation vote by the House Democratic caucus at the beginning of each Congress. In 1975 the caucus—mainly because of the infusion of younger, more liberal members in the "Watergate election" of 1974—used its new powers to depose three elderly southern committee chairs: Wright Patman (Texas) of the Banking Committee, W. R. Poage (Texas) of Agriculture, and F. Edward Hebert (Louisiana) of Armed Services, Hebert being replaced by the fourth-ranking Democrat on his committee.[8]

Thus by 1980 the old southern domination of the House was largely at an end because the old South itself was dead. The next section of this chapter looks at why southern Democratic House members remain affiliated with the modern Democratic party when the mainstream of that party outside the South appears to be so culturally and ideologically antipathetic to them.

Why Are They Still Democrats?

When asked why they are still Democrats, several of the southern Democratic House members made their initial response in terms of the nature of southern party politics during their formative years and at the time of their first involvement in political affairs. They were Democrats because their family and social milieu were Democratic and being a Democrat was one of the prerequisites of political success in that context. The following responses were typical:

Congressman Walter B. Jones (North Carolina): My part of the country has been predominantly Democratic until three or four years ago. Back when I was a child, if you were a Republican, people looked at you strangely.[9]

Congressman Earl Hutto (Florida): I guess because it was the traditional thing. In southern Alabama where I grew up, we didn't know what a Republican was. When I moved to Florida as an adult, it was pretty much the same. There were very few Republicans in north Florida, so I registered as a Democrat down there.[10]

Congressman Buddy Darden (Georgia): I'm a Democrat because my parents were and because Georgia is basically a Democratic state and the Democrats control all the political machinery.[11]

Congressman Butler Derrick (South Carolina): I'm a Democrat in the pragmatic sense because everybody else was a Democrat when I grew up in the South and started running for political office.[12]

Congressman Bob Clement (Tennessee): I wanted to eat. My father became governor when I was nine years of age and I needed a place to be and enough to eat. Since I was a young kid of nine years old, I knew we were Democrats from top to bottom.[13]

Only eight of the thirty members mentioned family and social milieu and being a Democrat as prerequisites of political success as their primary motivation, however, and only four of those gave no other reason. This reply was also more common among older members, whereas younger members tended to answer more in terms of issues or ideology, as opposed to family and regional tradition, perhaps a reflection of the transformation of the southern political universe over the past two decades or so.

In explaining their affiliation with the Democratic party, several of the House members that I interviewed made some mention of a concern for the economically underprivileged and the enhancement of social mobility:

Congressman Charles Bennett (Florida): I am a Democrat because the Democratic party set as its goal the uplifting of the opportunities of the underprivileged. And that is a noble goal.[14]

Congressman Owen Pickett (Virginia): The Democratic party tends to focus on human issues — education, health care, retirement — people who cannot handle issues on their own behalf. The government has to step in and provide more extensive relief.[15]

Congressman Glenn Browder (Alabama): I'm a Democrat because I think in the battle between the haves and the have-nots both sides are right, but the have-nots are the ones that need help. The haves can take care of themselves.[16]

Congressman Beryl F. Anthony, Jr. (Arkansas): I'm a Democrat because in my part of the country the Republican party was always represented by the most wealthy, country club–attending individuals . . . The Democrats were always the party of individual rights and liberties, of hope and opportunity for the common man.[17]

Congressman Claude Harris (Alabama): Although southern Democrats are fiscal conservatives, we also have a care and understanding of the plight of poor folks.[18]

Congresswoman Liz Patterson (South Carolina): The Democratic party has backed people programs like Social Security, and they finally passed Medicare, Aid to Education, and civil rights and human rights. The Democratic party is more of a people party; the other party is more of a business-oriented party.[19]

Congressman John Spratt (South Carolina): I think that the Democratic party genuinely believes in open opportunity and upward mobility.[20]

In explaining their affiliation, other members tended to emphasize the Democrats' latter-day commitment to civil rights:

Congressman Steve Neal (North Carolina): At the heart of our system is the idea that every individual is deserving of dignity and respect, regardless of sex or race, and the Democratic party reflects that better and tries to enhance the human condition better.[21]

Congressman Lewis Payne (Virginia): The Republicans in our state still stand for a certain status quo. Virginia had, even recently, a history of not very good race relations and didn't have a very progressive outlook on social issues. A number of us in the South feel that we've got to get beyond that.[22]

The commitment to civil rights obviously had special significance for the two black southern Democratic members whom I interviewed:

Congressman John Lewis (Georgia): In the late 1950s when we started the sit-ins, I was a student in Nashville and I began to identify with the Democratic party. Most of us had very little participation in the Democratic party or the Republican party, although there were pockets of black Republicans. I grew up in rural Alabama, and my family could not even register to vote.

But most blacks in the South already saw the Democratic party as the party of hope and opportunity. Our people fell in love with FDR. Black Americans had an allegiance to the Democratic party because FDR did something for black Americans.[23]

Congressman Mike Espy (Mississippi): I never had a choice. My family had always been Democratic, and the Republican party in the South has always been a narrowly based party and always inimical to the cause of full political and civil rights for black voters in Mississippi.[24]

In explaining their distinctiveness from the Republicans, several southern Democratic members also referred to a distinction between small business and big business, associating their party more with the interests of the former:

> *Congressman Ed Jenkins (Georgia)*: I'm a Democrat primarily by choice, in that all things being equal, when there's a distinction between big business and small business, the Democrats generally come down on the side of small business.[25]

> *Congressman Mike Parker (Mississippi)*: I'm a Democrat because I believe in the value of the working individual. I view it as what is good for small business is good for the health of the country. It's the backbone of the country. There's a large distinction between small business and big business. People approach business as being a monolith, but it's not true.[26]

> *Congressman Lewis Payne (Virginia)*: We all share concern for the small-business person. We want the little guy to have the opportunity to get into business.[27]

When asked whether they had ever contemplated switching parties, most of the interviewees dismissed the question fairly quickly with a negative response. In doing so, they frequently alluded not to issues or ideology but to the fact that the Democratic party is the majority party and is likely to stay that way in the House. The members thus would have nothing to gain by switching parties and a considerable amount to lose in terms of seniority and committee positions:

> *Congressman Buddy Darden (Georgia)*: I'm comfortable in the Democratic party majority. To be frank, I'd rather be in the minority within the majority, rather than in a majority within the minority. The Democratic party wouldn't be the majority party without us in the South. . . . I've never been tempted to switch. Most people don't trust people who switch parties.[28]

> *Congressman Jerry Huckaby (Louisiana)*: I always felt that I would be more effective here by being a member of the party that was in control. It was clear that the Democrats already had a hold on the House.[29]

Congressman Doug Barnard (Georgia): No, I've never been tempted to switch. . . . Now I'd lose my seniority and my subcommittee chairmanship. I've never been really sympathetic with partisan politics.[30]

Three of the more prominent southern conservative or "boll-weevil" Democrats emphasized the importance of maintaining a conservative wing in the House Democratic party and saw themselves as having an important moderating role:

Congressman Charles Stenholm (Texas): The difference between the parties is that the Democrats believe that there's a place for government. Our problem in the Democratic party is that we've let a very strong liberal minority take control. I want the Democrats to be a party with a social conscience but business common sense. The Republicans are antigovernment. They believe that no government is the solution. I strongly disagree with that. A strong vocal minority within the GOP controls their agenda, which tends to believe in no government.[31]

Congressman J. Marvin Leath (Texas): After I was elected to Congress in the early years of the so-called Reagan revolution, I had a lot of opportunities and pressures to change parties. I, being a practical practitioner of the political arts, didn't think that was necessarily the way to do it. I think that it's important that the Democratic party have a conservative influence in it to counterbalance the ultraliberal wing.[32]

Congressman Ed Jenkins (Georgia): The thought [switching] crosses most of our minds at some time in our lives, especially when we see presidential nominees that are so liberal that there's no hope of them carrying our districts or states. But those thoughts are somewhat fleeting. Most southern Democrats elect to stay within the party and fight it out.[33]

These arguments were not accepted by Congressman Andy Ireland (Florida), who successfully switched to the Republicans in 1984 after serving three terms as a southern Democrat:

I saw very clearly that changes in the committee structure after the onslaught of new liberal Democratic members in 1974 and 1976 had really changed the whole ball game. The old saw that southern conservatives balanced out policymaking by holding committee chairs was inaccurate. . . . The Democrats have entrenched this political system in Congress. They're a delivery group that supports a system that doesn't work.[34]

It is clear, however, that as long as the Democratic party retains a solid grip on the House of Representatives, only a handful of members who feel as ideologically uncomfortable with the Democratic majority as Congressman Ireland did are likely to contemplate switching to the GOP.

Issues and Ideology

Southern conservatives have traditionally been associated with reactionary positions across the entire spectrum of issues, especially race relations. As I already mentioned, however, this picture was not entirely accurate even in the era of the Solid South. Although compelled to adhere to the regional line on race, on other issues there were wide divisions among southern Democrats. In the late 1930s, although a group of southern Democrats joined in the famed "conservative coalition" to help thwart Roosevelt's New Deal, many others were enthusiastic supporters of Roosevelt, including the infamous Mississippi race baiter Theodore Bilbo![35]

In the contemporary House, the southerners remain ideologically distinctive from their fellow Democrats, although perhaps less so than in the recent past. The most dramatic turnabout on southern Democratic voting has, of course, been on issues related to race. This can be seen most dramatically in Table 4.2, showing the percentage of southern House Democrats who have supported the various voting rights acts since 1957. After the passage of the 1965 act, the gradual movement of southern black voters into the electorate had a transforming effect on southern Democratic voting patterns on civil rights issues in the House.[36] On race the South experienced a genuine partisan realignment, and the change in the Democratic position on civil rights was a manifestation of the fact that black voters now constituted the base of the Democratic vote in most southern elections.[37]

According to Congressman John Lewis, the increasing black political mobilization of the past two decades has not been lost on the white southern Democratic House members:

> One thing about white southerners is that they can count, and after the Voting Rights Act, they learned to count very well. . . . More and more black and white politicians get elected by biracial coalitions. You can no longer go into the black community and say one thing and then say something else to the white community if you are going to be elected and provide leadership.[38]

Table 4.2. House Southern Democrats' Support
for Voting Rights Legislation, 1957 to 1981

Year	Percent of Southern Democrats in Favor
1957	14
1960	7
1965	33
1970	34
1975	73
1981	91

Source: Adapted from Table 3 in Charles S. Bullock, III, "The
South in Congress: Power and Policy" in James F. Lea, ed.,
Contemporary Southern Politics (Baton Rouge: Louisiana
State University Press, 1988), pp. 177–93.

Lewis's analysis was supported by several of the white southern Democratic members from districts with substantial black populations:

Congressman Robin Tallon (South Carolina): In fact, in my district, the blacks are more politically involved than the whites are. . . . The core of black involvement is the black church. Sunday after Sunday you'll find me in black churches talking about politics. The church network is real strong. . . . They've always been politically active and involved at the grass roots.[39]

Congressman Lewis Payne (Virginia): We all share a pretty large black constituency, and we're interested in trying to do the right things to accommodate black views, on issues like the recent civil rights bill. In the first race I ran, blacks were very mobilized and were the reason I won. As long as they're all with me, it's hard for anyone else to win. I like to work with that constituency. I made sure black leaders and friends from my district got to see Nelson Mandela.[40]

Nevertheless, some members conceded that balancing the views of their black and white constituents was still a somewhat difficult, if not insurmountable, task:

Congressman Charles Hatcher (Georgia): If you represent a district in a strong two-party state like New Jersey or Florida, you expect a substantial challenge each time, and you'll have your own identifiable constituency out there that you represent. That's easier to represent than my district where I have to represent black voters – who are not liberal on all issues but tend to be more liberal – and very conservative whites, particularly on social issues. My supporters don't fit as neatly into a philosophical category.[41]

Others emphasized that the task of holding together their black–white electoral coalitions had been expedited by the effects of the southern electoral realignment that had moved the most race-conscious white voters to the Republicans:

Congressman Buddy Darden (Georgia): A lot of the racists went to the Republican party after the Goldwater campaign in 1964. That saved the Democratic party in the South and enabled them to build a coalition of moderate whites and blacks. Goldwater won a temporary victory, but he got the more racially sensitive elements out of the Democratic party and into the GOP.[42]

Congressman Jimmy Hayes (Louisiana): If someone wanted to move on the politics of race today, they'd register as a Republican. David Duke feels that the Democratic party is not the vehicle to use, because clearly the percentage of blacks in the Democratic party is greater.[43]

Although the difference on civil rights between the House Democrats as a whole and the southerners has all but disappeared, on other issues the southerners are still the most deviant group in the House Democratic party. This can been seen from Table 4.3, which shows important recent House votes on economic, cultural, and national security issues. On each of these dimensions the southern Democrats were clearly more conservative than the rest of the party.

The capital gains vote revived the Republican–southern Democratic alliance that had helped pass Ronald Reagan's tax cuts in 1981, after a period of virtually unprecedented party unity among House Democrats in the late 1980s. According to Congressman Ed Jenkins, who led the southern Democratic revolt on behalf of the Bush administration proposal:

I've been concerned with capital gains since the Jenkins–Steiger resolution of 1978. I had to lead the fight on that as a freshman. I believe in it and I've seen it work. I don't believe in class warfare. It would be a tremendous mistake for our party to try and attack high-to middle-income people on issues like this. . . . There's some truth of the charge that higher-income people have gotten a big break during the Reagan era. But I could make the

Table 4.3. Southern Democratic Deviations on Three Key House
Votes, 1989 to 1991 (percents)

Vote	All Democrats in Favor	Nonsouthern Democrats in Favor	Southern Democrats in Favor
Capital gains tax reduction 9/29/89	24	14	52
Flag desecration amendment 6/21/90	37	27	61
Authorization of force in the Persian Gulf 1/12/91	31	18	65

Sources: *New York Times, 29 September 1989; New York Times,* 14 January 1991; and
Congressional Quarterly Weekly Report, 21 June 1990.

tax code a lot more progressive than it is today. . . . It would hit some
higher-income people but not do it in an area where they're trying to attract
venture capital.[44]

When asked on which particular issues they departed most from the party
leadership, most of the members that I interviewed first mentioned
economic issues, particularly legislation such as child care and family
leave, which mandates businesses to provide certain benefits to their
employees. This discomfort with "mandated benefits" reflects the South's
status as a developing economy and also the southern Democratic members' widespread empathy for small businesses.

Congressman Mike Parker (Mississippi): The federal government's job is
not to mandate every corner of our lives but to make sure that an atmosphere is created so that people can go into business and get a decent job and
have hope for the future.[45]

Congressman John Spratt (South Carolina): I'm less inclined to be for
legislation like family leave. As a lawyer I'm a little concerned about
the litigiousness of our society, and I'm wary of creating laws that make
us more litigious. The average Democrat is more willing to engage in
social engineering. My views are partly based on instinct and partly my
constituency.[46]

Congressman Doug Barnard (Georgia): The liberal mainstream feels that the government should participate in every solution for problems in the community, even for problems that don't necessarily exist. Take the child-care bill, for example. In my area this is adequately served by the churches and private enterprise. . . . We believe that the solution should begin in local communities, the states, and private enterprise. Only in matters of great magnitude should the federal government become involved.[47]

A few members also mentioned that the support of the party mainstream for the positions held by organized labor—which remains a weak political force in most of the South—made them feel particularly uncomfortable:

Congressman Steve Neal (North Carolina): Democrats are too inclined to use government excessively . . . on labor issues. Sometimes these are power issues that deal with giving more power to labor than I think is healthy.[48]

Congressman Lewis Payne (Virginia): Generally our constituents are not too much in favor of unions and the causes they're espousing. Virginia is a "right-to-work" state, and I'm proud of that. It's not a matter of not identifying with working people but of not identifying with the organization that claims to be speaking about working people. They've become rather self-serving.[49]

Many southern Democrats nevertheless retained a "populist" tone in their remarks on economic issues and were eager to demonstrate their empathy for the have-nots or the common man. In practice this translated into perhaps a greater sensitivity than the party mainstream had to the plight of small business and middle- to low-income people, who felt themselves squeezed economically by the tax and benefits system:

Congressman Jimmy Hayes (Louisiana): Neither political party has been able to appeal to families who make too much money to quality for any programs but too little to put their kids through school. Moynihan showed that middle America is overtaxed. Who's paying for the homeless and the S&L cleanup? Middle America. I can't think of anything that's been done for those people in the last twenty years, but I can think of plenty that's been done *to* them.[50]

The distinctiveness of the southerners also persisted on foreign and defense policy. Given the southern military tradition and the large contri-bution of defense to the southern economy, southern members remained more supportive of an assertive American foreign policy and far less skeptical with regard to the merits of American intervention overseas than did their Democratic brethren in the rest of the United States. Surprisingly, given their voting records, not one of the thirty southern Democratic members that I interviewed mentioned defense when asked on which issues they felt most at odds with the party mainstream. When prompted, several southern members were anxious to stress their commitment to a strong defense:

> *Congressman Earl Hutto (Florida)*: I differ on that with many of my liberal friends. It's in our best interests from a security standpoint, and in north Florida, it brings so much to the economy. It's a good government expenditure that provides jobs while at the same time it assures our security.[51]

Members were even less eager to mention the highly sensitive moral and cultural issues such as abortion, gay rights, school prayer, and freedom of speech, although when pressed, most of them admitted to a far more conservative position than that of their nonsouthern fellow Democrats. Southern Democratic members obviously felt rather vulnerable on moral and cultural issues because these issues forced them to choose between the traditional values of their constituents and the mainstream liberal viewpoint of their party colleagues:

> *Congressman Glenn Browder (Alabama)*: I also differ on traditional values—symbolic things like the flag, patriotism, and family values. Those symbols and concepts rate higher for me.[52]

> *Congressman Bob Clement (Tennessee)*: I want us to hold onto our traditions and values. I'm proud of the South and its heritage and tradition. . . . It's unbelievable the way people are rising in the South in terms of jobs and opportunities, but I hope we don't forget about the importance of the values that made the South very distinctive.[53]

> *Congressman Doug Barnard (Georgia)*: Southerners seems to be more patriotic and more loyal to moral issues than those who are more supportive of rights and freedom of expression and choice. These things divide the Democratic party, and old-fashioned values are more apparent among southern Democrats.[54]

> *Congressman Owen B. Pickett (Virginia)*: Southerners disagree on many issues, but they tend to have views that are more traditional. In the South, families have been there years and years, for generations. Family members

fought in the revolutionary war and the Civil War and in World War I and World War II, so that their values and thinking about the country are based on family, history, and tradition.[55]

According to black Mississippi Congressman Mike Espy, southern concern for traditional values was equally shared by both his black and white constituents: "I have opposed gun control and supported prayer in schools and the death penalty because I believe they are reasonable value issues. I am Mississippian. I was born and raised there, and I believe in what Mississippians believe in."[56]

From this discussion of the positions of the contemporary southern Democratic House members, we can see that according to most indices the southern members have come closer to the Democratic party mainstream. On civil rights issues in particular, the old southern distinctiveness has disappeared. As Congressman David Price observed: "My impression is that across the South, members are much less conservative and are a much less distinctive bloc than we were twenty years ago. . . . We're no longer wearing white linen suits and filibustering civil rights bills."[57]

Yet from my examination of voting patterns in the House and from conversations with southern members, it became clear that despite the civil rights revolution and the end of the Solid South, the southern Democrats were still the most distinctive bloc in the party, and the bloc that was most likely to deviate from the Democratic norm across the entire spectrum of issues, with the exception of civil rights. The next section examines the southern Democrats' relationship with the party leaders and the degree of influence they were able to exert in the modern House relative to that of their forefathers.

Southern Democrats and the House Democratic Party

Although still distinctive from the rest of the House Democratic caucus, southern Democrats have come much closer to the Democratic mainstream than they were a quarter-century ago.[58] Table 4.4 shows the party unity scores of the southern Democrats, which have risen markedly since the 1960s, to a point that in the late 1980s they voted as often with their party as the Republican members did! Table 4.5 shows that the conservative coalition between Republicans and southern Democrats that controlled the House during the 1940s and 1950s appeared only very infrequently in the House in the late 1980s, although when it did appear it had a very high success rate.

Table 4.4. Party Unity Scores in the House, 1970 to 1988 (percent)

Year	All Democrats	Southern Democrats	Republicans
1970	71	52	72
1971	72	48	76
1972	70	44	76
1973	75	55	74
1974	72	51	71
1975	75	53	78
1976	75	52	75
1977	74	55	77
1978	71	53	77
1979	75	60	79
1980	78	64	79
1981	75	57	80
1982	77	62	76
1983	82	67	80
1984	81	68	77
1985	86	76	80
1986	86	76	76
1987	88	78	79
1988	88	81	80

Note: Table shows the percentage of members voting with a majority of their party on party unity votes (i.e., votes on which a majority of a party votes on one side of the issue and a majority of the other party votes on the other side).

Source: Norman J. Ornstein, Thomas E. Mann, and Michael J. Malbin, *Vital Statistics on Congress, 1989–90* (Washington, DC: American Enterprise Institute, 1990), p. 199.

It is interesting that since the enforced retirement of House Speaker Jim Wright in June 1989, there has been no southerner on the House leadership team. In the leadership elections that month, Richard Gephardt (Missouri) defeated Ed Jenkins (Georgia) by 181 votes to 76, and Beryl Anthony gained only 30 votes in losing to Bill Gray (Pennsylvania) in the race for party whip.[59] Although neither Congressman Jenkins nor Congressman Anthony believed that it was impossible for a southern Democrat to be elected to a leadership position in the present-day House, they both conceded that it had become much more difficult than previously, given the decline in the southern proportion of the caucus:

Congressman Ed Jenkins (Georgia): It is a problem for the party when you have four liberals in the top four positions. Generally, on a lot of issues and

Table 4.5. Conservative Coalition Appearances and Victories in the House, 1970 to 1988

Year	Appearances (%)[a]	Victories (%)
1970	17	70
1971	31	79
1972	25	79
1973	25	67
1974	22	67
1975	28	52
1976	17	59
1977	22	60
1978	20	57
1979	21	73
1980	16	67
1981	21	88
1982	16	78
1983	18	71
1984	14	72
1985	13	84
1986	11	78
1987	9	88
1988	8	82

[a]The percentage of all votes on which a majority of southern Democrats and a majority of Republicans opposed a majority of northern Democrats.

Source: Norman J. Ornstein, Thomas E. Mann, and Michael J. Malbin, *Vital Statistics on Congress, 1989–90* (Washington, DC: American Enterprise Institute, 1990), p. 200.

personal requests, they are responsive, but it's difficult for a southern member to enter the leadership. Before they changed the rule that the whip was elected, they used to use that opportunity to balance the leadership. . . . Jim Wright was a southerner and he ran and was elected, but he wasn't very conservative. It's difficult for a southern member to be elected to either of the top two positions, and they don't want to offer for those positions because they'd get thirty to forty votes and be embarrassed.[60]

Congressman Beryl Anthony (Arkansas): [Running for the leadership] probably forces you to be much more liberal than what your area or state requires. There's a serious balancing act between trying to develop respect and reputation in Washington, DC, and not losing it in your home area. I do think that it's tougher for southerners in that regard. Not impossible, but tougher.[61]

Congressman Jimmy Hayes (Louisiana) also pointed out, however, that a particular kind of southerner could aspire to a leadership position:

> My background and voting record would prevent me from holding a position in the party leadership, but remember there are pockets within the South that are quite different to my district. For example, Hale Boggs got to be majority leader because he came from an urban district. . . . When Dick Gephardt beat Ed Jenkins, Jenkins lost because he was perceived as a rural southern Democrat. A southern Democrat by that definition cannot hold a leadership position.[62]

It is not surprising, therefore, that despite the apparent greater ideological convergence of southern and national Democrats in the House, the question of their relationship with the party leaders provoked contrasting responses from southern members. Although most agreed that the leadership was willing to give them a hearing, three relatively junior southern Democrats felt that it was not particularly responsive to their concerns:

> I have a feeling sometimes, and others have it too, that we don't fare as well as others have because the leadership thinks we're not real Democrats.[63]

> The present leadership is not responsive to southerners. But you must remember that I'm only in my second term and I don't have much occasion to deal with them.[64]

> The party leadership listens, and that's it.[65]

By contrast, other members emphasized that the Foley–Gephardt–Gray leadership was more accommodating to the southern Democrats than its predecessors had been:

> *Congressman Jerry Huckaby (Louisiana)*: The leadership has changed since the beginning of the boll weevils. O'Neill and Wright have both gone. . . . Now they try to recognize and accommodate the southern Democrats.[66]

> *Congressman Steve Neal (North Carolina)*: I don't find the leadership to be unfair, and I think they allow all positions to be debated.[67]

> *Congressman John Spratt (South Carolina)*: The Democratic party in the House is not tightly disciplined, and the leadership does not whip us in any real way. The whip count is more count than whip. When there are local or regional reasons for our votes, they are very understanding, but accom-

modation is expected on important procedural votes like the committee choices, caucus decisions, and voting for the rules and decisions of the chair — votes that help run the business of the House.[68]

Congressman Earl Hutto (Florida): The party leadership is understanding of my situation. They know I have a very conservative electorate despite the high percentage of Democratic registration. In my belief, they would rather have a Democrat, regardless of philosophy, than a Republican to represent the district.[69]

The leadership's control over committee assignments is probably the most significant power that it wields over the membership of the caucus. In this respect it is clear that the leadership does tend to reward and punish members for their cooperation and that several southern members felt somewhat disappointed with the assignments they had been given:

On assignments I'm not sure. I've come to the conclusion that I'll never be Speaker of the House or majority leader, since my voting record is not consistently Democratic.[70]

I'm not going to get any choice assignments on Ways and Means or Appropriations, although there are a significant number of southern Democrats on those committees.[71]

I've been reminded on occasion that when I told the leadership I couldn't support them, it could have an effect on my committee assignments and reassignments.[72]

In fact, a glance at Table 4.6 reveals that the southern Democrats, though not as dominant as they were in the 1950s, still held a disproportionate share of the House committee chairmanships in 1990. In regard to committee assignments, it is also interesting that southern Democrats still tended to gravitate toward the same committees (see Table 4.7). The percentage of southern Democrats on the Armed Services Committee was essentially the same as it was in 1970, and Table 4.8 indicates that the Democrats on that committee remained substantially more conservative than those on most other committees. Southern Democrats were still disproportionately represented on the Agriculture and Veterans Affairs committees, although the percentage in the former case had fallen from 63 percent in 1970. On the three major committees — Appropriations, Rules, and Ways and Means — the southern proportion fell from a slight overrepresentation to a figure proportionate to their share of the House caucus.

Table 4.6. House Committee Chairmanships Held by Southerners, 1955 to 1991

Year	Percent of Democrats from the South	Percent of Committee Chairs from the South	Percent of Three Major Committee Chairs from the South [a]
1955	43	63	67
1967	35	50	100
1975	28	28	33
1981	29	27	33
1991	29	40	33

[a] Ways and Means, Rules, and Appropriations.

Source: Norman J. Ornstein, Thomas E. Mann, and Michael J. Malbin, *Vital Statistics on Congress, 1989–90* (Washington, DC: American Enterprise Institute, 1990), p. 123; and Michael Barone and Grant Ujifusa, *The Almanac of American Politics 1992* (Washington, DC: National Journal, Inc., 1991), pp. 1455–76.

Table 4.7. Southern Democratic Representation on House Committees, 1970 and 1990

Committee	Percent of Southern Democrats, 1970	Percent of Southern Democrats, 1990
All Democrats	33	30
Agriculture	63	44
Appropriations	33	29
Armed Services	47	48
Banking	33	23
District of Columbia	50	0
Education and Labor	0	0
Energy and Commerce	na	27
Foreign Affairs	14	27
Government Operations	25	21
House Administration	29	23
Interior	11	12
Judiciary	25	19
Merchant Marine	24	35
Post Office	20	7
Public Works and Transportation	31	29
Rules	40	33
Science	40	33
Standards (Ethics)	50	0
Veterans	43	43
Ways and Means	40	30

Source: *Congressional Quarterly Almanacs 1970 and 1991* (Washington, DC: Congressional Quarterly Press, 1971 and 1992), p. 35D, p. 61D.

Table 4.8. Democratic House Committee Members' Support for the Conservative Coalition, 1959 and 1987

Committee	1959	1987
Chamber Average	**44**	**41**
Agriculture	69	56
Appropriations	55	41
Armed Services	55	63
Banking	26	39
Budget	na	33
District of Columbia	65	17
Education and Labor	16	28
Energy and Commerce	49	36
Foreign Affairs	32	28
Government Operations	37	40
House Administration	52	38
Interior	36	36
Judiciary	45	26
Merchant Marine	39	46
Post Office and Civil Service	40	17
Public Works and Transportation	45	42
Rules	44	35
Science and Technology	34	52
Small Business	na	48
Standards (Ethics)	na	38
Veterans Affairs	46	52
Ways and Means	42	36

Source: Norman J. Ornstein, Thomas E. Mann, and Michael J. Malbin, *Vital Statistics on Congress, 1989–90* (Washington, DC: American Enterprise Institute, 1990), pp. 201–3.

The most dramatic fall occurred on the District of Columbia Committee, which was dominated by conservative southerners in the 1960s owing to fear that the capital, with its heavy black population, might be the thin end of the wedge on desegregation if allowed self-government. After the civil rights revolution, southerners completely lost interest in this committee. (Southerners have also disappeared from the Ethics Committee, but because this is an assignment that members conspicuously seek to avoid, it should probably not be perceived as indicating a diminution of their power). Besides, the District of Columbia Committee, the committees that held little interest for the southerners were the same as in 1970—Education and Labor (not a single southern Democrat on this committee in either year!), Foreign Affairs, and Interior.

On the ideological scale, as measured by support for the conservative coalition, the committees favored by the southern Democrats in both 1959 and 1987 were those with the most conservative Democratic memberships in the House. The three major committees became significantly, but not overwhelmingly, more liberal than they had been in the late 1950s. The Judiciary Committee became much more liberal because of the civil rights revolution, and with the disappearance of the southerners, the District of Columbia Committee fell from having the second most conservative Democratic membership in the House to having the most liberal Democratic membership in 1987. Education and Labor, shunned by southern Democrats in both years, remained one of the most liberal committees in the House.

The most significant change, then, was that southern conservative influence on the three major committees of the House had markedly decreased and that the Democratic membership of those committees had become more representative of the Democratic caucus as a whole. Probably more significant than the proportional decrease, however, is the reduction of the chairman's control over the committees and the easing of the seniority rule. Thus, even when a southerner ascended to the chairmanship of an important House committee—such as Jamie Whitten (Mississippi) of Appropriations or Walter Jones (North Carolina) of Merchant Marine—they exercised nothing like the degree of control that their southern forebears did. Moreover, given that committee chairmanships were now subject to a possible vote in the full caucus, both these veteran Democrats had been compelled to move their personal voting position closer to that of the House Democratic mainstream. According to Congressman Jones:

> When I first got here, my voting pattern was extremely conservative. For a few years, I got an annual award from the Americans for Constitutional Action. . . . In my particular case, I assumed the chairmanship of a committee in 1980, and that changes your pattern of voting. I now tend to vote more with the leadership.[73]

In short, the southern House Democrats suffered a noticeable decrease in their influence because of the southern electoral realignment and changes in the House power structure. However, they were by no means in as weak a position in the contemporary House Democratic caucus as several of them appeared to believe. The leadership was well aware of the need to secure their support and so did pay attention to their concerns.

Complaints about treatment by the leadership tended to be more prevalent among junior members lacking significant experience with the House's power structure. My interviews with other members indicated that the leadership could be very accommodating to southern Democrats who made a particular effort to approach them:

> *Congressman John Spratt (South Carolina)*: I told the Speaker he needed a core of leadership supporters on the budget committee, and within three weeks a seat opened up. In that sense he was very responsive. . . . One problem that we have is that we tend to be moderate and centrist, and so we aren't quite as vocal or obstreperous as the people to the right and left of us. We're not as aggressive, and we've only ourselves to blame for not getting the attention . . . and it helps us not to have too much of a regional identification.[74]

> *Congressman Charles Stenholm (Texas)*: Tip [O'Neill] eventually became very responsive. I remember one meeting we had with him when things were going rough within the caucus. He reminded us that a squeaking wheel gets grease and we weren't squeaking enough. After that we got one conservative member in his advisory group.[75]

The leadership also took particular care to use some members as a conduit to the southern Democrats. During the 1980s Congressman Beryl Anthony (chair of the Democratic Congressional Campaign Committee from 1986 to 1990) and Congressman Ed Jenkins both played this valuable role for the leadership:

> *Congressman Beryl Anthony (Arkansas)*: I have an informal but not a formal role. On most critical issues when they want to get the true feeling of the southerners, I serve as a go-between. But it's a strictly informal relationship.[76]

> *Congressman Ed Jenkins (Georgia)*: It has been true to some extent. They [the leadership] make the assumption that if I cannot support a measure, then it is going to be difficult to pass. If I can't support it, those more conservative than I won't support it.[77]

Without the southern Democrats the party leadership's majority would have disappeared, and being aware of this fact the leadership, perforce, had to pay attention to southern Democratic opinion. Southern Democrats, on the other hand, were well aware of the situation and tried to achieve the most that they possibly could from their crucial bargaining position, as we shall see in the following section.

Regional Identity and Organization

Most of the members that I interviewed tended to agree that a southern identity still existed but that it was much weaker than it had been in the 1940s and 1950s, when the southern Democrats rallied around the segregation issue. In the current House, owing to the impact of reapportionment and the Voting Rights Act, the southern Democratic members are a much more diverse group than they were fifty years ago, and even then they were not monolithic. According to veteran Florida Democrat Charles Bennett: "There's a tendency to categorize all southerners who wash their face and keep their clothes clean as "old southern gentlemen." But the truth is that there is just as wide a variety of members as in the North. There are very few cotton farmers in my district."[78] Georgia Congressman Charles Hatcher agreed: "It's not really like it was decades ago. We're not a monolithic group. Southerners representing urban areas are quite liberal on most issues. It's not a bloc like it was twenty–thirty years ago. The *Baker* and *Brown* decisions changed that."[79]

It appears that southern members still tend to organize on particular issues that they have in common, and they still gravitate more toward certain committees as opposed to others. With the emergence of the Republican party in the South, however, the southern Democratic members find that when they act "regionally" these days, they must do so in a bipartisan fashion:

> *Congressman Mike Parker (Mississippi)*: According to what the issue is, certain bonds pull us together. One is agriculture, another is the economic climate. For example, 63 percent of the people covered by the minimum wage are from the eleven states of the southeast region.[80]

> *Congressman Steve Neal (North Carolina)*: There are some issues where we work more closely, for instance, agricultural issues, but even there it's broader than the South. I work a lot with people from my state, but that's nonideological, and then I work with both Democrats and Republicans.[81]

> *Congresswoman Liz Patterson (North Carolina)*: Regional matters tend to be bipartisan—as in the "Sunbelt caucus"—because so often our region does not get the dollars.[82]

Congressman Charles Hatcher (Georgia): We talk among ourselves and share ideas more than with people from other regions, but we don't necessarily strictly organize along those lines.[83]

Even though the Sunbelt caucus is bipartisan and includes members from western states, a conservative group specifically for Democrats has existed since 1980. The Conservative Democratic Forum (CDF) is not a specifically southern organization, but according to one of its members, the active membership is "90 percent southerners."[84] The catalyst for the founding of the CDF was Ronald Reagan's 1981 budget, when the group organized the so-called boll-weevil Democrats in support of the Republican plan.[85] Nevertheless, sentiment for conservative Democrats to organize themselves and to exploit their balance-of-power position had been increasing during the late 1970s. According to the group's founder, Texas Congressman Charles Stenholm:

> I formed the CDF group with the goal of providing a forum for moderate-to-conservative Democrats to discuss the issues before Congress in an attempt to form a consensus and affect the direction of the party. . . . We also wanted to get a more conservative balance on key committees — especially Budget, because on some key conservative issues, a position that comes out of committee is very hard to defeat on the House floor.[86]

Stenholm's Texas Democratic colleague Marvin Leath stressed the importance of the 1980 election results in providing the boll weevils with their opportunity:

> We agreed with a lot of the things Ronald Reagan said he wanted to do. We agreed that the tax system needed to be reformed, that our defense effort needed to be strengthened, and that the Great Society programs should be cut back and eliminated.
> We saw the thirty-five conservative Democrats held the balance of power in the House, and we thought it would be a good idea to have a loosely linked group of people to try and make things happen.[87]

The budget of 1981 saw the revival of the conservative coalition with a vengeance. On Reagan's 30 percent income-tax cut, thirty-six of the forty-eight Democrats who supported him were southerners.[88] Much of the budget package had been put together by Democratic Congressman Phil Gramm (Texas) and Reagan's budget director, David Stockman, and the leverage of the CDF and its members was at its height. But the onset of recession and the Democratic gains in the 1982 midterm elections put an end to the boll weevils' balance-of-power position. After being thrown off the Budget Committee by a vengeful Speaker O'Neill, Gramm defected to the GOP. Most of the fiscally conservative members of the CDF, however, became progressively disenchanted with the Reagan administration, owing to Reagan's apparent lack of concern for the ballooning federal budget deficit. According to Congressman Leath:

> In the last five or six years, the Reagan presidency became very nonresponsive to political realities, particularly regarding the budget deficit. He cut taxes too much in 1981 and increased defense spending too quickly. Also, in 1980/81 Reagan was able to deliver close on 100 percent of the Republicans, which wasn't the case after that.[89]

The CDF did not disappear after 1982. In 1990 it had forty-six or forty-seven members, meeting once a week on Thursday for about twenty minutes. The group has no staff of its own, and no congressional staffers are permitted to attend the meetings. Some of these meetings are "face-offs" in which the CDF members hear representatives of two conflicting viewpoints on a topical issue or pending piece of legislation and ask questions. Members also use the group to get cosponsors for legislation.[90] Chairman Stenholm emphasized, however, that the CDF is a forum and not a coalition or a caucus, with the organization and discipline implied by a caucus, and that it seeks to cooperate with the leadership rather than set itself up against it:

> We never even tried to be a coalition. Out of forty-six or forty-seven of us, we usually split twenty-two to twenty-three, depending on the issue. We were never a group marching in lockstep to any one political idea. . . . The major difference between us and the [House moderate Republican] Wednesday Group is that we're in the majority and they're in the minority. We, therefore, have to work with the leadership in the majority party.[91]

Virtually all of the thirty southern Democratic members that I interviewed were CDF members, although they varied in both their commitment to the group and the degree to which they still found it to be useful. The less-active members tended to emphasize time pressures and the fact that the group was no longer so important to them because it had lost its balance-of-power role:

Congressman Charles Hatcher (Georgia): I'm peripherally involved. I was just here and naturally I joined. . . . I think that as a group, it's not nearly as active as it was in those first couple of years, when we made the difference for Reagan. Now I don't find it really useful.[92]

Congressman Robin Tallon (South Carolina): I was associated with the CDF and I may still be on the roll, though I haven't been to a meeting in two years. It's a good group of people, but there are too many commitments on my time for me to be active. It's become more fractured in the last couple of years.[93]

Some other of the less-involved members also expressed reservations about being perceived as too conservative by associating too much with the CDF:

Congressman Jimmy Hayes (Louisiana): I don't want the conservative Democrats to be viewed as too far right. I think of myself as a moderate. I'm very conservative on fiscal issues, but I'm liberal on social [welfare] issues.[94]

Congressman Ed Jenkins (Georgia): Some southerners are also not comfortable meeting with archconservative members, because they will not vote that way. Their districts are not such that they are able to vote for the right wing, so they don't come. Very few of the members of our delegation—perhaps only one or two members—seem to come very often.[95]

Those members who still get some benefit from their CDF membership tended to emphasize the importance of being able to discuss issues with like-minded party colleagues and to provide at least some minimal degree of organization of the conservative Democratic forces in the House:

Congressman Lewis Payne (Virginia): It's useful in terms of looking ahead at the concerns that will come up during the session.[96]

Congressman Jerry Huckaby (Louisiana): The CDF meets weekly, and from time to time we use it as an educational forum where we have the different sides of issues presented to us.[97]

Congressman Tim Valentine (North Carolina): It's the only influence within the Democratic party here that says "slow down."[98]

Congressman Marvin Leath (Texas): It's still good to have a loose-knit organization to get together and occasionally talk about things.[99]

Some other CDF members maintained that the group also served a useful function as a channel of communication to the Democratic leadership:

> *Congressman Earl Hutto (Florida)*: The bond between us has helped quite a lot in dealing with the leadership, because we have a considerable bloc of votes. We sometimes can be the swing vote for Democratic proposals or Republican proposals. It's been very useful.[100]

> *Congressman Claude Harris (Alabama)*: We're not always together on different issues, but it's a numbers game up here and everybody understands numbers. . . . We hope to get the leadership to understand that we've got a different view sometimes on legislation, and it gives us an opportunity to provide input.[101]

The CDF appears increasingly to be a group waiting for another opportunity to play the balance-of-power role in the House that they played in 1981. In the interim it is serving the function of keeping the southern and conservative Democratic forces in communication, although it takes no votes or stands on legislation and does not seriously try to organize its members. The CDF has forced the party leadership to pay some more attention to their southern conservatives. Like their moderate Republican counterparts, however, the conservative Democrats may have suffered from a disinclination to organize aggressively on a regular basis, as have their more ideologically oriented brethren in the House Democratic caucus.

Southern House Democrats and the National Party

During the 1980s the southern Democratic House members were remarkably successful in getting easily reelected in districts that were generally carried equally easily by Republican presidential candidates Ronald Reagan and George Bush.[102] It was the South that exemplified the "split-level realignment" of the 1970s and 1980s better than any other region did, and it was clear that for most southern voters there were two Democratic parties: the one that they encountered in House elections, with which they still felt comfortable, and the one exemplified by the party's presidential candidates in 1984 and 1988, from which they appeared to be totally alienated. The southern House members had become masters of the art of differentiating between themselves and the party's presidential nominees. Alabama Congressman Glenn Browder, who won a 1989 special election in a district where George Bush won 60 percent of the vote in 1988, explained the phenomenon in this way:

> The Republicans do very well controlling their message and symbols at the national level. Translating that into victories in specific geographical districts is more difficult. . . . Mass constituencies are more responsive to the manipulation of symbols like "liberalism," the "flag," or "Willie Horton." When you get down to constituencies where the delivery of services is concerned, people are more amenable to those who can deliver. You have different levels of conceptualization. At the top you get ideological conceptualization; at the lower levels you have group conceptualization; that is, "What does it mean to a person like me?"[103]

Congressman Mike Espy of Mississippi, who has been consistently reelected in a very racially divided district, concurred:

> The decline of the Democrats in presidential elections has everything to do with personality. In congressional elections our success has everything to do with ability to use the tools of incumbency such as the frank and constituent service. A lot of us feel we don't owe a lot to the national party and try to dissociate ourselves from the national ticket. We choose to run on local issues.[104]

Some members even admitted that it better suited their own electoral purposes to have a Republican in the White House, since they feared that having a Democratic president might compel them to adhere more closely to a party "line" in the House and might associate them too much with a national Democratic party that is unpopular in the South:

> I feel secure in my relationship with the Bush administration. I can play a position as a broker between the Democratic leadership and the administration. I would not be as important to a Democratic administration, because I would have to defend them and begin to take litmus tests. I'm more important with George Bush in the White House than with Al Gore.[105]

Most of the members that I interviewed would not go so far as to say they actually preferred dealing with a Republican president (many of them had never known anything else!), but they expressed strong preferences for a Democrat from their own region, as opposed to another northeastern liberal. The Democrats' 1988 presidential nominee Michael Dukakis created a particularly bad impression among the southern Democratic House members:

> *Congressman Mike Parker (Mississippi)*: The Dukakis people came to Mississippi and thought they could tell us how people in Mississippi thought about issues. They thought they knew better than we did, and when they found out what Mississippians really did think, they wrote us off.
>
> The concept is this: Dukakis could not get me one vote in Mississippi, but I could have gotten Dukakis votes. But when you disagree with everything the sucker says, how can you turn around and support him?
>
> . . . In 1988 my opponent tried to tie me to Dukakis, but my granddaddy told me, "If you try to carry two watermelons at the same time, you're gonna bust both of them." My watermelon was my own election.[106]
>
> *Congressman Charles Bennett (Florida)*: I supported Mr. Dukakis and all the Democratic nominees to the best of my ability, but the things that are important to my constituents are not important to these candidates. . . . There has been a lack of attention to the symbolism of the South. Southerners expect candidates to recover themselves better than Mr. Dukakis did. When the question of the pledge of allegiance and the flag came up, Dukakis should have responded aggressively. The importance of the symbolism did not occur to him.[107]

When asked to explain why the Democrats had put up nominees who were anathema to the South, virtually all the members blamed the party's nominating process, arguing that the latter gave too much importance to interest groups, the national news media, and small northern states such as Iowa and New Hampshire, none of which had much in common with the concerns of most southern Democrats:

> *Congressman Doug Barnard (Georgia)*: Our nominating process is structured against southern conservative candidates. . . . Iowa has no more votes than I have in my congressional district, and New Hampshire and other isolated areas do not reflect the general support of the Democrats. . . . The Democrats are dominated by special interest groups, and they do not relate to one another or to the populace. There's no hope that you're going to change the situation until we start giving fair representation of the entire Democratic populace.[108]
>
> *Congressman Jimmy Hayes (Louisiana)*: The Democratic party's selection process does not reflect anything. . . . An eight-hour voting experience

in high school gyms to begin the process is not reflective of the way a person plays the game. Iowa Democrats are way more liberal than the national average, and the Republican party has a large number of the religious far right. How many people do you know who would vote for eight hours? Relying on extreme activists is not the way to do the selection process.

. . . The only time we won was when Carter went around the mechanism and not through it. He beat them, so they changed the process to make sure that it didn't happen again. And they succeeded. It hasn't happened since.[109]

When pressed, most members admitted that they had no ready solution to the problem. Many supported the moderate Democratic Leadership Council (DLC) (see Chapter 5) in its aspiration to change the ideological direction of the party, although the DLC is not specifically committed to reforming the nominating process. One member (who requested anonymity) suggested that for the process to be changed in a direction more favorable to southern conservatives, the whole primary system might have to be abandoned.

The DLC wants to form organizations to elect people as delegates to the DNC (Democratic National Committee) or the convention, because mainstream leaders don't participate in the DNC or the convention any more. Why should I run as a delegate in my district when I might lose? The DLC's hidden agenda is that if the Democrats get wiped out one hundred to zero in a presidential election, they want to organize the states. They want to get control of the DNC so that you no longer have a process that ends up with a plurality candidate like Dukakis or Jesse Jackson.[110]

Despite their remarkable success in averting a Republican takeover at the congressional level in the South, the southern Democratic members in the House were fearful of the long-term consequences of continued Republican domination at the presidential level. Nevertheless, few if any of them were willing to take a prominent role in trying to change the whole structure of the system in a direction more favorable to more moderate or conservative Democrats like themselves. Moreover, many of them readily conceded that having a Republican in the White House placed them in a more advantageous position electorally in their districts and strategically in the House. So although there was a fear of a presidential electoral calamity so great that it would actually begin to move large

numbers of southern Democratic seats to the Republican column, most
members were confident that they could continue to divorce their own
electoral context from the national context as effectively as they had done
in the 1980s.

Summary

The civil rights revolution and the electoral realignment that it engendered finally ended the southern domination of the House of Representatives in the early 1970s. Nevertheless, southern Democrats remained a
substantial presence in the House. They still controlled about two-thirds
of the southern House districts, and they still exerted an influence
disproportionate to their numbers in the chamber because of their relative
sense of regional identity and cohesiveness and their balance-of-power
position between the parties. The House leadership learned from the
experiences of 1981 that alienating the southern conservatives too much
might be very costly in terms of control over the House.

At the same time, however, much had changed. Southern Democrats
were now virtually unanimous in their support for civil rights, and the
effects of reapportionment and black voting rights had created a greater
degree of representativeness and diversity in the southern Democratic
ranks. This "liberalizing" effect was, of course, reinforced by the movement of the most conservative sections of the southern states to the
Republican party.

The southern Democrats were no longer such a clearly deviant group
in the national Democratic party as they once had been; they had come
much closer to the party mainstream. Yet as long as the South remained
the most distinctive region of the United States and as long as it continued
to elect a preponderance of Democrats to the U.S. House, the southern
Democrats would likely remain a distinctive and significant element in
the Democratic ranks.

The Senate

The "Solid South" in the U.S. Senate

In the "Solid South" era the U.S. Senate was the strongest bastion of
southern influence in the American federal government. Although the
southern Democratic House members were able to exploit the committee

system and the seniority rule to enhance their power, their influence over the chamber as a whole paled beside that of the southern Democratic senators (see Table 4.9). When William S. White and Donald Matthews wrote their contrasting accounts of the workings of the U.S. Senate in the 1950s, one item on which both agreed was the extent that "southern" mores had virtually come to define the Senate as an institution. The "folkways" of the Senate—the deference to senior members, the elaborate courtesies, the institutional pride—all were specifically southern characteristics, especially as enunciated by such representatives of the tradition as Senators Harry F. Byrd, Sr., of Virginia and Richard B. Russell of Georgia.[111]

The South had essentially taken over the Senate, for the same reasons that had increased southern influence in the House: the impact of the specialized committee system with its autocratic chairmen, and the operations of the seniority rule. In the Senate, however, southern influence was strengthened by several other features specific to the institution. The first of these was the more individualistic character of the Senate. Whereas the House with 435 members requires something of a hierarchical internal power structure to enable any business to get done, the Senate with only 100 members gives far more power to the individual member. There being no Rules Committee in the Senate, a single senator can introduce bills on the floor and submit nongermane amendments or "riders" to proposed legislation.

Table 4.9. Democratic Senate Seats by Region, 1924 to 1988

Region	1924	1936	1948	1960	1972	1980	1988
South and Border	66	66	56	43	35	40	38
Plains and Midwest	2	18	7	15	23	19	24
New England and Middle Atlantic	10	16	11	11	16	21	20
Mountain and Pacific	22	24	26	31	26	19	18
	100%	100	100	100	100	100	100

Source: Norman J. Ornstein, Thomas E. Mann, and Michael J. Malbin, *Vital Statistics on Congress, 1989–90* (Washington, DC: American Enterprise Institute, 1990), pp. 15–16.

The Senate also gave the southern senators the weapon of the filibuster. With no effective Rules Committee to schedule legislation, debate in the Senate ended because of either the individual members' sense of fair play or a vote of two-thirds of the total membership. The southern Democrats, with some border-state allies, were generally able to summon a blocking third on any legislation they deemed inimical to the interests of the region, and the southerners' experience and seniority also gave them a mastery of Senate rules and procedures.[112]

The southern bloc in the Senate was further distinguished by the advent of the conservative coalition, with the Republicans opposing Roosevelt's court-packing plan of 1927 and the further expansion of the New Deal.[113] For a period of some thirty years the coalition effectively ruled Congress and thereby restricted the growth of the domestic state and precluded any erosion of segregation and disfranchisement in the South. In 1947 the coalition achieved probably its greatest triumph by passing the Taft–Hartley Act, which greatly restricted the ability of labor unions to organize in states — like those in the South — that adopted right-to-work laws. Given the party balance of the Senate during most of this period, the southern Democratic senators were the key to the passage or failure of most legislation, and they exercised this leverage to the maximum possible extent on behalf of their states. According to Patterson, writing about the leaders of the original southern conservative revolt against the New Deal in the 1930s:

> So long as the New Deal did not disturb southern agricultural, industrial or racial patterns, these leaders would support it, sometimes with enthusiasm. But if and when the northern wing of the party began to dominate (as it did after the 1936 election) a certain degree of friction was almost inevitable.[114]

At the time that White and Matthews were writing about the Senate's "inner club" during the 1950s, there could be little doubt that most of the members of that club were southern. My interviews with the surviving southern senators of that era revealed the particular influence of two or three of their colleagues. According to former U.S. Senator Russell B. Long of Louisiana:

> In 1948 [when Long first entered the Senate] the South had a lot of influence. There was a gentleman's club, and Dick Russell, Harry Byrd, and Walter George were the key players. They were very talented people and gentlemen in every meaning of the word. In the main they subscribed

to a gentleman's code, and they supported it. They would never ask a direct question, but it was implied that there was a condition of reciprocity, that they might have something they had to ask of you.[115]

Former Senator J. William Fulbright of Arkansas concurred in assessing the impact of Richard Russell:

> Dick Russell from Georgia was by much the most important senator. . . .
> Russell was a very fine man. He was a bachelor, and so he had no diversion
> from a family. He gave his undivided attention to the Senate. The seniority
> rule was respected, and the juniors deferred to a much greater degree than
> they do now to senior senators.[116]

During the 1950s the southern Democratic senators, led by Richard Russell, constituted the last line of defense against federal legislation that would dismantle the southern social and political system. In general they did their job well. By the time that President Kennedy took office in 1961, only two very attenuated civil rights bills had been passed through the Senate, in 1957 and 1960, and although the Supreme Court had directly attacked segregation by reversing the *Plessey* decision in the celebrated case of *Brown v. Board of Education*, the decision was generally ignored and not enforced throughout the South.[117] Yet in the early 1960s the Solid South was crumbling rapidly, and when it fell, the rationale for southern Democratic dominance in the Senate disappeared as well.

A New Senate: Ideology and Partisanship

Since the days of the "inner club" and southern Democratic domination during the mid-1950s, the U.S. Senate has been transformed. By the 1980s even more power had devolved to individual senators, and the notions of an inner club and deference toward seniors had been replaced by a more blatantly partisan and ideological approach. The election of the large class of liberal Democratic senators in 1958 and the selection of Mike Mansfield as Lyndon Johnson's successor for the post of majority leader in 1961 marked the beginning of this process.

The thirteen new Democratic senators after the 1958 elections were predominantly liberal and activist in their approach to government. Having been elected in largely Republican states on the basis of their personal appeal, they were no longer so willing to "go along in order to get along" as the more traditional Senate incumbents had been and to tolerate the autocratic rule of southern committee chairmen who were ideologically far to the right of the Democratic caucus as a whole. Mansfield's approach to leadership in the Senate was to regard himself as the servant of the individual senators rather than as the "leader of the Senate," and his passive style of leadership accorded perfectly with the requirements of the new class of liberal Democrats.[118]

An additional factor was the final passage of the Civil Rights Act in 1964 and the Voting Rights Act in 1965, which finally ended segregation and the disfranchisement of blacks. When this finally happened, the power of the southern bloc in the Senate was broken, and the whole *raison d'être* of southern solidarity was undermined. Senator Russell Long contrasted the positions taken by the southern senators before and after the 1964/65 revolution:

> The southern members had little choice. In my case, if I had taken a position contrary to that the southern group was taking on civil rights, I would have been defeated for office. One reason was that so few blacks were voting in the southern states at that point. When I first went to the Senate [in 1948] there were only ten thousand black voters. When I ran six years later, there might have been eighty thousand registered, but only about half of them turned out.
> . . . When they passed the Civil Rights Law of 1964, it made a big difference. I saw what was happening. Something had to happen, and on the whole I was pleased to see it. . . . After the Voting Rights Act, there were a lot more black voters, and that made a big difference in the way southern legislators would vote and the attitude they'd take toward measures. All of them voted against the Bork nomination except Hollings.[119]

The new, more liberal Senate came about largely because of the lopsidedly liberal Democratic victories in the Senate elections between 1958 and 1978. Together with the small band of progressive Republicans, such as Jacob Javits of New York and Clifford Case of New Jersey, the Democratic liberals could now fairly consistently outvote the conservative coalition.[120] The transformation was also assisted by the advent of a "New South" generation of politicians in the wake of the Voting Rights Act. Although most of them would never have described themselves as liberals, they held progressive views on civil rights and sought to build black–white electoral coalitions.[121] By the mid-1970s, their success in this endeavor had reduced the segregationist core of southern Democrats to a small number of aging senators.

Table 4.10. Party Unity Scores in the Senate, 1970 to 1988

Year	All Democrats	Southern Democrats	Republicans
1970	71	49	71
1971	74	56	75
1972	72	43	73
1973	79	52	74
1974	72	41	68
1975	76	48	71
1976	74	46	72
1977	72	48	75
1978	75	54	66
1979	76	62	73
1980	76	64	74
1981	77	64	85
1982	76	62	80
1983	76	70	79
1984	75	61	83
1985	79	68	81
1986	74	59	80
1987	85	80	78
1988	85	78	74

Note: Table shows percentage of members voting with a majority of their party on party unity votes (i.e., votes on which a majority of a party votes on one side of the issue and a majority of the other party votes on the other side).

Source: Norman J. Ornstein, Thomas E. Mann, and Michael J. Malbin, *Vital Statistics on Congress, 1989–90* (Washington, DC: American Enterprise Institute, 1990), p. 199.

The Senate also became a more partisan institution as the ideological polarization between the Republican and Democratic parties increased. Although the House, with its peculiar constituency orientation, remained somewhat immune to this kind of pressure, the Senate, with its higher profile on issues—especially foreign policy, social and cultural issues—and its greater national visibility, became something of an ideological battleground. Judicial confirmations, in particular, provoked bitter partisan struggles and a resort to tactics by senators and their interest-group allies that would have been unthinkable two decades earlier. The competitiveness of Senate elections (partisan control of the chamber changed twice during the 1980s), in contrast with House and presidential contests, also served to raise the partisan stakes (see Table 4.10).[122]

Table 4.11. Conservative Coalition Appearances and Victories in the Senate,
1970 to 1988

Year	Appearances (% of votes)[a]	Victories (%)
1970	26	64
1971	28	86
1972	29	63
1973	21	54
1974	30	54
1975	28	48
1976	26	58
1977	29	74
1978	23	46
1979	18	65
1980	20	75
1981	21	95
1982	20	90
1983	12	89
1984	17	94
1985	16	93
1986	20	93
1987	8	100
1988	10	97

[a]The percentage of all roll call votes on which a majority of southern Democrats and a majority of Republicans opposed a majority of northern Democrats.

Source: Norman J. Ornstein, Thomas E. Mann, and Michael J. Malbin, *Vital Statistics on Congress, 1989–90* (Washington, DC: American Enterprise Institute, 1990), p. 200.

In this new partisan atmosphere, southern Democrats have drawn much closer to their northern brethren in their voting records. With the dropping of the old barrier of segregation, the senatorial Democratic party has now become a much more consistently liberal group, and the senatorial Republican party, much more consistently conservative, because senatorial politics has become more strident and ideologized. Yet even though they were generally well to the left of southern Republicans, the southern Democrats were still a distinctively "moderate" group in the senatorial Democratic caucus, and with the chamber quite finely balanced between the parties during the 1970s and 1980s, they remained in a crucial "swing" position. Most critical votes in the contemporary Senate are decided by the votes of the southern conservative (relative to the rest of the Senate caucus) Democrats (see Table 4.11).

Senate Southern Democrats in the 1990s

When we look at the profile of southern Senate Democrats in the 1990s, we can see that many of the old stereotypes no longer apply. The greatest change, of course, has been on the race issue (see Table 4.12). With the realignment of southern politics, most of the anti–civil rights sentiment transferred to the GOP, and the southern Democratic party became a biracial alliance of blacks, feminists, teachers, labor unions, and other groups associated with Democratic party politics nationally. The difference in the South, however, is that the aforementioned groups are numerically much weaker, so that for statewide victories, Democratic Senate candidates still needed a substantial vote from lower-status white voters – the old George Wallace constituency – who had defected in large numbers to the GOP in presidential politics. During the 1970s and 1980s, through skillful campaigning, New South Democrats were remarkably successful in pulling off this feat.

In contrast with the House, there are no formal or institutional groups in the Senate in which the southern Democrats can gather as they did in the 1950s. Part of the reason for this is institutional: In the House, individual members can gain more clout by participating in groups and caucuses, whereas in the Senate the individual senator only has to be recognized by the presiding officer to be heard. According to Louisiana Senator John B. Breaux, who served in both chambers during the 1980s:

Table 4.12. The Southern Democratic Realignment on Civil Rights

Category	Percent Voting for 1964 Civil Rights Act	Percent Voting for 1991 Civil Rights Act
Senate:		
All Democrats	69	100
Nonsouthern Democrats	98	100
Southern Democrats	5	100
House:		
All Democrats	63	98
Nonsouthern Democrats	95	98
Southern Democrats	8	99

Sources: Charles and Barbara Whalen, *The Longest Debate: A Legislative History of the 1964 Civil Rights Act* (New York: Mentor Books, 1985), pp. 247–54; and *Congressional Quarterly Almanac, 1991* (Washington, DC: Congressional Quarterly Press, 1992), p. 31S, p. 94H.

It has to do with the nature of the differences between the institutions. In the House you have a large number of southern Democrats. In the Senate it's two-four-six-eight and we're done. The nature of the group is much smaller.

The pattern in the House is to have a caucus for every issue. The Senate not only does not have a southern caucus, it's not in the nature of the Senate. The time requirements are more severe than in the House, and I represent eight times more people than I did in the House.

It's unusual to have a Democratic southern caucus. We do confer, talk, and speak to each other on how bills affect our area, but we don't do it in any formal sense.[123]

Senator Richard Shelby of Alabama concurred: "We confirm people in the Senate. It's a smaller body, and we get involved in different things than in the House. It's hard to rank and rate senators. All that is very misleading."[124] Tony Garrett, press secretary to Louisiana's senior senator, J. Bennett Johnston, emphasized the lack of coordination among contemporary southern senators: "On some issues you see it, but there's less of that than there used to be. It's not the days of Stennis and Kerr anymore."[125]

The absence of formal linkages among modern southern Democrats in the Senate is a mark of the extent to which regional cohesion has decreased in the post–Voting Rights Act era and perhaps of the increasing economic, social, and political diversity of the southern states. These days southern Democrats are no more cohesive than are Democrats from other regions, joining forces only when some common regional concern is at stake.

When asked to explain why they are Democrats, most southern senators, like their House counterparts, respond in terms of family and upbringing. The economic populist tone prevalent among the House members also could be heard among the senators and staffers:

Senator Richard Shelby: The Democratic party is big enough and broad enough to have a lot of us of various persuasions, and I hope it will continue to be. Being a Democrat meant support for a lot of programs that have provided equal opportunity and small business advocacy. It used to mean support for a strong national defense.[126]

Senator John Breaux: The Democratic party best represents the ideas and goals of middle-class working people: people who need the help and assistance of government. The Republicans believe in a trickle-down formula, where the very wealthy do well. I don't think that works. If working-class people do well and succeed, they buy goods and services. So I think a "trickle-up" theory works better than trickle-down. I grew up as a traditional Democrat, and I've felt more comfortable about that as I've been exposed to the philosophy of the Republican party.[127]

Blaine Bull, legislative director to Senator Lloyd Bentsen of Texas: Bentsen got into the Democratic party because it was the only party in a one-party state. He's always talked about being a Democrat in terms of the diversity of the party, while the Texas Republican party remains fairly narrow in its values. He sees himself as a moderate Democrat. Conservatives see him as conservative. Moderates see him as moderate, and liberals see him as moderate. He's conservative on some issues and liberal on others.[128]

On virtually every important vote after the Democrats regained control of the Senate in 1986, it was the southern Democrats who decided the issue. The first was the nomination of Judge Robert Bork to the U.S. Supreme Court in 1987. The Bork nomination provides a clear illustration of the contradictory pressures on modern southern Democratic senators and how they resolve them. Conservative pressure on behalf of Bork was counteracted by a powerful campaign from black leaders and civil rights groups against the nomination. As it transpired, the pressure from the anti-Bork forces was greater than that from the conservative camp, and one by one the southern Democratic senators came down against the nominee, with the exception of Senator David Boren of Oklahoma and Senator Ernest Hollings of South Carolina.[129] Aside from feeling personally uncomfortable with Bork, the southern senators had to face the new reality of Democratic party politics in the South: The party's base of support lay in the black community. To alienate that base was to endanger their chances of reelection. This reality was brought home to the other southern senators by Senator J. Bennett Johnston of Louisiana. According to his press secretary:

A number of senators were talking about Bork, and somebody asked Senator Johnston how he was going to vote. He told them, "I don't know how I'm going to vote, but I know how you're going to vote" and reiterated to them some factors about politics in their own states.[130]

Johnston's Louisiana colleague John Breaux described the general southern Democratic discomfort with Bork:

It was a personal thing. I didn't make the decision based on qualifications. Judge Bork on some issues was further outside the mainstream than people think. He was an agnostic, which was not in keeping with what I would like to see. I felt that he was outside the mainstream and that what was needed was a moderate conservative. Bork was more ideological than I think a judge needs to be. There was also strong opposition to him within the state.[131]

Southern Democrats again played a significant role in the next great Senate confirmation battle, in 1989, over the nomination of ex-Senator John Tower to be secretary of defense. In this instance, the crucial role was played by Armed Services Committee chairman Sam Nunn, who on defense issues had established an authority among his fellow southern senators almost equivalent to that of his Georgia predecessor, Richard Russell. In the face of Nunn's opposition, only two southern senators — Howell Heflin of Alabama and Lloyd Bentsen of Texas — supported the doomed nominee.

As in the House, the southern senators emerged as the most dissident element in the Democratic party in 1990/91 on the proposed constitutional amendment to prohibit desecration of the American flag and in support of giving President George Bush the authority to wage war on Iraq (over the opposition of Sam Nunn). In the fall of 1991 there was a replay of the Bork nomination, only on this occasion the Supreme Court nominee was a black conservative, Clarence Thomas. Thomas's race defused potential opposition from the black political community in the southern states, and it was clear that a comfortable majority of southern senators were going to support his confirmation, even though the southern Democratic representative on the Senate Judiciary committee, Senator Heflin of Alabama, was opposed. Although the accusations of sexual harassment by Thomas's former aide, Anita Hill, somewhat eroded Thomas's southern support, enough southern Democrats supported the nominee to guarantee his narrow confirmation (see Table 4.13).

The recurrent pattern in these votes is that on issues pertaining to national defense and social and cultural issues (but excluding civil rights), southern Democrats in the Senate were still somewhat to the right of the mainstream of their party, both in the Senate and nationally. When I asked the southern senators or their close aides about the areas in which they differed from the mainstream of the party, their responses tended to focus on defense and social issues:

Table 4.13. Southern Democratic Deviations on Three Key Senate Votes, 1990 and 1991 (percents)

Vote	All Democrats in Favor	Nonsouthern Democrats in Favor	Southern Democrats in Favor
Flag desecration amendment 6/26/90	36	28	60
Authorization of force in the Persian Gulf 1/12/91	18	10	40
Clarence Thomas confirmation 10/15/91	19	10	47

Sources: Congressional Quarterly Almanacs 1990 and 1991 (Washington, DC: Congressional Quarterly Press, 1991 and 1992), p. 2S, p. 29S, p. 30S.

Senator Richard Shelby (Alabama): On some issues I'm pretty conservative, and on some others I haven't been so conservative. I try to represent the mainstream Alabama moderate-to-conservative, but not liberal, voter.[132]

James Fleming, administrative assistant to Senator Wendell Ford (Kentucky): He's most conservative on social issues – abortion, the death penalty, and school prayer. In Kentucky these issues are more positive than not.[133]

Blaine Bull, legislative director to Senator Lloyd Bentsen (Texas): He's always believed in prayer in school, and he's always been for the balanced budget. He's never been a big proponent of an activist federal judiciary. These are all rather conservative points of view.[134]

Tony Garrett, press secretary to Senator J. Bennett Johnston (Louisiana): He has been more prodefense than is reflected by the national party posture. He's somewhat to the conservative side on social issues like abortion. He supported the flag amendment.[135]

These positions are the product of the southern states' persistent adherence to more traditional social values and an instinctive prodefense posture. The ability of the southern Democratic senators to reflect those values while remaining Democrats explains their ability to survive electorally during the 1980s while Democratic presidential candidates tumbled to defeat throughout the region.

Again like their House counterparts, the southern Democrats in the Senate became exasperated by the dismal performance of their party's presidential candidates in the South. More so than in the House, the minds of the Senate's southern Democrats were concentrated on the need to transform the nominating process so as to make it more representative of their constituents' views. Senator Terry Sanford (North Carolina) had been particularly prominent in arguing over the years for changes in the Democrats' nominating process, even to the point of running himself (unsuccessfully) in 1985 for chair of the Democratic National Committee. According to Sanford's longtime associate Sam Poole:

> After the 1968 convention, we changed the process so that it's impossible to get anyone nominated who can get elected. The delegates to the national convention don't represent the people; they represent candidates. . . . In his book *The Danger of Democracy* Sanford argued that we had to get back to the process of selecting delegates at the local level and let the delegates select the candidates instead of the candidate's selecting the delegates. . . . If by some fluke, we elect a Democratic president, that would make it easier to change the process. Carter had the chance, but his attitude was "Why should I change the process that nominated me?"[136]

Senator John Breaux favored a system of regional primaries:

> It will change when people get tired of losing and when we get the message across that unless we get middle-class working people back in the Democratic column, we'll continue to be noncompetitive with regard to the presidency.
>
> We could change the process by shortening the length of the campaign and by having regional primaries which raise regional issues as opposed to the narrow issues that have plagued us in the general election.[137]

All of the senators whose views are represented here were supporters of the Democratic Leadership Council (DLC), though only Senator Breaux had taken a leading role in that organization.

As individual members, senators have a greater degree of clout than do congressmen. They are not so dependent on good relations with the party leadership to secure good committee assignments, and they are much less subject to any kind of attempt by the leadership to discipline them or apply sanctions. Nevertheless, the leadership still organizes the Senate's schedule, and amicable relations are important to the southern members.

During the 1980s, discontent with the leadership of Senator Robert Byrd (West Virginia) led to challenges by Senator Lawton Chiles (Florida) in 1984 and Senator Bennett Johnston (Louisiana) in 1986. In both cases the challenge petered out before a formal vote was taken, but the senators did clearly indicate some discomfort with the direction of the leadership. According to Senator Wendell Ford's (Kentucky) aide James Fleming:

> The leadership changed to a degree, and rank-and-file Democrats weren't brought in on anything. Byrd held everything close to his chest, although Ford and Byrd were close personally. Part of the thing was that the leadership wasn't doing as much as it ought to be for Democratic senators. It wasn't any philosophical thing.[138]

In 1988 Bennett Johnston was a major contender for the position of majority leader when Byrd resigned to take over the chair of the Appropriations Committee, but he was easily defeated by the liberal northerner George Mitchell of Maine. Although the issues here again were not strictly regional or ideological, Johnston's aide Tony Garrett believes that Johnston's southern conservative background was certainly more of a liability than an asset: "That's difficult to say. The votes hinged so much on one-to-one relationships and were more or less automatic. For those whose votes hinged on nonpersonal relationships, things like political philosophy could have been a factor."[139]

It is clear from these comments, however, that the job of senate leader is so much more a matter of personal relationships than it is in the more hierarchical House and that regional and ideological considerations appear to be strictly secondary. Southern associations did not prevent Senator Ford from securing the Senate whip's position in 1991.

Summary

Certainly the loss of southern Democratic identity seemed to be greater in the Senate than in the House. This is partly due to the fact that the impact of the Voting Rights Act was probably greater at this level. Although there still are many homogeneous southern white rural or semiurban congressional districts, the southern states now have much more diverse electorates for their senators to represent, with substantial minority constituencies. This has been reflected in the pattern of recent Senate voting on civil rights legislation and in Supreme Court nominations. When asked why his voting record had become markedly less conservative when he moved from the House to the Senate, Senator John Breaux responded:

> It's a reflection of the fact that I now represent eight districts. My congressional district was more conservative than the state at large. It's one thing to represent one district, but to represent an entire state is different. You have different voters and concerns, and you have a different perspective. You have a larger area to represent and be responsive to.[140]

The civil rights revolution also had a major impact on the distinctiveness of the southern Democrats in the Senate, by removing the *raison d'être* of that distinctiveness, the race question. The need for the southerners to organize to defeat civil rights was no longer present, and a new generation of southern Democratic senators emerged who were more sympathetic to the goals of the national Democrats. Urbanization and the communications revolution have also served to reduce regional pressures on southerners. From being a distinct entity in the Senate in the 1950s, southern Democrats have become more like a regional group similar to those of senators from other regions of the country; that is, they coalesce on an ad hoc basis around issues of interest to their region.

The smaller size of the Senate, the power and respect accorded to individual senators, and the general contempt for organized groups also have reduced the need for caucuses to coordinate members as they do in the House. In the Senate there is no need for a Conservative Democratic Forum on the House's lines. According to Senator Bennett Johnston's aide Tony Garrett, what regional and ideological coordination there is in the Senate "is loose knit and nothing formal. It's something that develops spontaneously issue by issue."[141]

Yet the persistent (if more muted) regional distinctiveness of the South does mean that there is likely to be more of a common voting pattern among southern Senators, especially on issues that impinge on central elements of southern political culture—the military, national security, respect for national symbols, sexual mores, and church/state issues—we have seen that the southern senators take a more conservative position.

Moreover, despite all that has changed since the civil rights revolution, one important fact remains as relevant today as it was in the 1950s: When Democrats control the Senate, the southern Democrats hold the balance of power between the GOP and their northern brethren, although they have come much closer to the latter in their voting patterns over the past thirty years.

5

The Conservative Counterattack:
The Democratic Leadership Council

In the discussion of American party factionalism in Chapter 1, I noted that one of the most salient features of the contemporary American party system has been the weakness of the ideological "center" in both of the major parties. The Downsian model of party competition prevalent during the late 1950s assumed that parties operated as unified teams and that in a two-party system the candidates and policies of the parties would inevitably gravitate toward the center of the ideological spectrum (where most of the voting public was situated).[1] Yet since the Republicans' nomination of Barry Goldwater in 1964, the major American parties have generally adopted presidential candidates and platforms more representative of the views of the parties' ideological activists than of the parties' identifiers in the general electorate.

The major reason for the recent weakness of the centrist or "moderate" wings of the Democratic and Republican parties is the transformation of American party politics that has been in progress since the Progressive era at the turn of the century and that culminated in the party reform movement of the 1968–72 period. Whereas traditional American party politics was based on ethnic or regional loyalties and was principally oriented toward securing political offices for the material benefits that party officeholders could confer on party supporters, the creation of a new bureaucratic elite selected according to merit and the rise of a larger professional middle class led to the supplanting of the old-style party "boss" by the idealistic "amateur" activist in both major parties.[2] A further consequence of this revolution was that the presidential (and, to a lesser extent, the congressional) Democratic and Republican parties now reflected the views of committed Democratic and Republican party activists (from the New Left and neoliberal factions in the case of the Democrats) rather than those of Republican and Democratic identifiers in the electorate.[3] The concurrent phenomenon of persistently divided government at the federal level can be partially explained as the rational response of a less ideologically committed electorate to the more polarized national Democratic and Republican party elites.

Each major party nevertheless retained a moderate or centrist faction (primarily southern in the case of the Democrats) that remained deeply skeptical regarding the polarized ideological politics of the "new presidential elites" and continued to gravitate toward the center of the political spectrum.

In an earlier work I analyzed the efforts of moderate or progressive Republicans to resist the increasing domination of the GOP by ideological conservatives in the post-1964 period.[4] I concluded that although "moderate Republicanism" was kept alive by regional and local "political cultures" in certain sections of the United States, the attempt by the Ripon Society (and various other groups) to reconstitute a vibrant centrist wing of the Republican party on a nationwide basis was doomed to failure. A successful presidential nominating campaign, traditionally a matter of mobilizing state and local party organizations and officeholders, now required the mobilization of single-issue interest groups and ideologically committed activists. This transformation precluded a national-level comeback by the liberal Republicans after they lost control of the national GOP in 1964.[5]

In this chapter I apply the same analysis to the Democratic Leadership Council (DLC), founded largely by southern Democratic moderates in 1985 and by far the most conspicuous of the moderate Democratic party groups of recent years. Interestingly, there are several parallels between the DLC's evolution and that of the Ripon Society in the late 1960s. Like the Ripon Society, the DLC did not lack support from party officeholders, favorable national attention, and detailed policy prescriptions. The DLC initially also shared several of the Ripon Society's most conspicuous weaknesses: too much of an elitist Washington orientation, no national fund-raising and campaigning apparatus, and the inadequacy of "moderation" as a means of mobilizing the activist cadre required for a successful national nominating campaign in contemporary America.

Yet in stark contrast with the Ripon Society, the DLC accomplished its major goal: As a self-proclaimed southern moderate and former DLC chair, Arkansas Governor Bill Clinton secured both the Democratic nomination and the presidency in 1992. The reasons for Clinton's success and whether that success portends the end of the "ideological" polarization between the American parties will be discussed in the next chapter and the conclusion. This chapter looks at the first phase of the DLC (1985–90), its organizational structure, and why—despite an auspicious beginning—the DLC was relegated to the sidelines during the 1988 presidential campaign.

The First Phase of the DLC

The DLC was founded in the wake of the Democrats' landslide defeat in the 1984 presidential election, but according to the DLC's executive director, Alvin From, the first stirrings of the organization began at the Democrats' 1984 convention in San Francisco:

> In 1984 [Virginia Governor Charles] Robb, as chairman of the Democratic governors, started to work with [Georgia Senator Sam] Nunn. At the San Francisco convention there was an effort by [former Vice-President and Democratic Presidential Nominee Walter] Mondale to reduce the number of superdelegates. We had a meeting in Gillis Long's suite at the Fairmont, which a lot of people think of as the beginning of the DLC.[6]

The extent of the Democrats' defeat in November gave further impetus to the effort to create a forum for primarily southern, moderate-to-conservative Democrats. Governor Robb became involved in a failed effort to prevent the election of Paul G. Kirk—a former aide to Senator Edward Kennedy—as chairman of the Democratic National Committee. This brought Robb together with a group of like-minded Democratic officeholders—Congressman Richard Gephardt of Missouri (who became the first chairman of the DLC), Governor Bruce Babbitt of Arizona, Senator Nunn, and Florida Senator Lawton Chiles—and they were the principal movers in the formal launch of the DLC in March 1985.[7]

The main motivation of these officeholders was the fear that the weakness of the party's liberal national candidates would eventually damage more conservative Democrats like themselves at the lower levels of electoral competition. The defeats of moderate candidates like Governor James Hunt of North Carolina and Senator Walter Huddleston of

Kentucky in the 1984 Senate races, and the considerable Republican pressure on southern Democratic officeholders to switch parties, concentrated the minds of conservative Democratic officeholders on the task of presenting an alternative face of the national Democratic party to that seen at the San Francisco convention in 1984.[8]

According to the DLC leadership, the major problem with the Democratic party in presidential elections was its association with a narrow range of interest groups, such as organized labor and militant feminist and minority organizations, rather than the broader concerns of the middle-class voters who had moved over to the Republican column during the 1980s. According to Senator Robb: "We plan to talk about growth, opportunity, strength, social justice and traditional values, which are important if we are to recapture the support of the American people."[9] When asked if he conceived the role of the DLC as being similar to that of the Ripon Society in the Republican party, Al From, the DLC's executive director since its inception, disagreed:

> The better Republican parallel might be the conservative movement in the Republican party and the Heritage Foundation. What Heritage was trying to do was to change the party intellectually. They knew that unless there was an intellectual renaissance, there was not going to be a political renaissance. Their challenge was to make conservative policies acceptable to middle-class Americans, where conservative policies were beyond the pale.
>
> We're trying to make progressive policies once again desirable to middle-class Americans. It's quite a challenge.[10]

Since its inception, then, the DLC has seen its mission as developing new themes and ideas for the Democratic party, in order to dilute the influence of the New Left ideology—referred to by the DLC as "liberal fundamentalism"—on the Democratic party.[11]

In pursuit of this goal, the DLC initially concentrated on recruiting support among the party's elected officeholders. The roster of DLC supporters among leading Democrats was impressive, extending from the expected southern and western senators, congressmen, and governors to more surprising individuals such as the black congressmen William Gray (Pennsylvania), John Lewis (Georgia), and Mike Espy (Mississippi). Charles Robb, who took over the DLC chairmanship from Gephardt in 1986, described his specific efforts to try and broaden the DLC's base:

It took a year or two for us to get credibility outside the South and West. . . . I wanted to broaden our geographic and racial base, so I got a number of big city mayors like Tom Bradley [Los Angeles], Harold Washington [Chicago], and Henry Cisneros [San Antonio] to join. That gave us a little more depth and showed that we weren't just a separated part of the party.[12]

Robb's efforts did not allay the persistent criticism of the DLC by the party's establishment, which viewed its activities with a great deal of suspicion. The relationship between the DLC and the party's national committee was particularly tense, since the DLC's founding fathers had been opposed to Paul Kirk's election as national party chairman in 1985. Kirk directly challenged the DLC's aspiration to be the fount of new ideas in the party, by establishing his own "Democratic Policy Commission," headed by former Utah Governor Scott Matheson, to debate Democratic policy positions on a variety of issues.[13] The DLC leadership remained mistrustful of the national committee as a bastion of "liberal fundamentalism." Yet despite the institutional rivalries, Kirk's view of the party's future direction did not differ essentially from that of the DLC, as was demonstrated when he abolished the party's "midterm conference" (which he viewed as a media showcase for militant single-issue activists) and the national committee's caucuses for women and minorities. Kirk further ensured that the so-called superdelegate slots for elected officials and party leaders were retained for the 1988 convention.[14]

Despite the favorable notices from the political media and its broad base of support among party officeholders, the DLC was a peripheral actor in the presidential politics of 1988 (see Chapter 3). Part of the problem was the absence from the field of the two potential contenders most associated with the DLC, Sam Nunn and Charles Robb, who could not be tempted into the fray despite the creation of the southern "Super Tuesday" primary on 8 March 1988. On the other hand, three leading members of the DLC were in the Democratic presidential field in 1988: Richard Gephardt, Bruce Babbitt, and Senator Albert Gore (Tennessee) as the champion of the South. None survived beyond the New York primary, leaving the nomination as a contest between Rev. Jesse Jackson (representing the New Left/minorities faction) and the neoliberal Massachusetts governor, Michael Dukakis.

Al From acknowledged the DLC's lack of impact in 1988:

One of the consequences of the [nominating] system we have now is that it forces everyone to play interest-group politics. Candidates without big

money have to play a different kind of politics, and I don't know what
would have happened if they had had the real ability to raise big money and
be competitive in every way with Dukakis. Dukakis was the one guy in that
race who could raise a lot of money.

We were only four years old and had no infrastructure and fewer
defining ideas, and our movement was a lot less of a national effort than we
are now.[15]

The absence of an effective grass-roots political organization or network
of groups to mobilize the moderate-to-conservative Democratic vote
ensured the victory of the liberal fundamentalists. Interestingly, the
dynamics of contemporary presidential politics also forced to the left
the supposedly moderate Democratic candidates. The Gephardt who ran
for president on an almost explicitly protectionist platform in 1988 hardly
seemed like the moderate Democrat who had been the first chairman of
the DLC in 1985. And it was perfectly rational for Gephardt to adopt a
more liberal tone, because all the evidence from the polls indicated that
the "Reagan Democrats," which the DLC was trying to bring back into
the Democratic party fold, no longer participated in large numbers in
Democratic party presidential primaries. The premise of Super Tuesday
(for which the DLC was blamed, although it had little to do with it) was
that it would provide a bloc of conservative Democratic delegates from
the South, but in truth the Democratic presidential primary electorate
south of the Mason–Dixon line these days is not vastly different from that
in other regions of the United States.

In 1988, the structure and environment of Democratic presidential
nominating politics, as Al From noted, rendered the DLC and its impressive roster of officeholders more or less irrelevant to presidential politics.
Primary elections—north or south—are won and lost by mobilizing
militant single-issue or ideological activists, a game in which Jesse
Jackson and Michael Dukakis had distinct advantages over the candidates
whose positions were closer to that of the DLC. Nevertheless, the extent
of Dukakis's defeat in November at the hands of the rather uninspiring
Republican George Bush was something of a vindication of the DLC's
viewpoint: Another liberal Democratic nominee had been defeated because his values regarding a variety of cultural issues were at odds with
those of a large number of formerly Democratic, white, middle-class
voters.[16] The DLC thus appeared to have an opportunity to transform a
very disappointing election into a much more influential position for
themselves within the Democratic party.

The DLC's Infrastructure

1990 the DLC's headquarters were on Capitol Hill, at 316 Pennsylvania Avenue, Washington, DC. The same building also housed the Progressive Policy Institute (PPI) — the DLC's "in-house" think tank — which was kept legally separate from the DLC for tax purposes. Each organization had a full-time staff of about ten persons.[17] The overall membership of the council included over four hundred elected officials (with members in all fifty states), and since 1991 Louisiana Senator John B. Breaux has served as the national chairman of the DLC (previous holders of the post have included Richard Gephardt, Sam Nunn, Charles Robb, and Bill Clinton). At the DLC's 1990 national conference in New Orleans, the then chairman, Arkansas Governor Bill Clinton, set the goal of establishing DLC chapters in every state, and by May 1991 over twenty states had established DLC chapters. Since September 1989, the DLC has published a bimonthly magazine (edited — until he left to work full time for Clinton's 1992 presidential campaign — by the DLC's policy director, Bruce Reed), originally entitled *The Mainstream Democrat* and including articles by DLC and PPI staffers, elected officeholders, and sympathetic academics and policy specialists. The title of the magazine was changed in 1990 to *The New Democrat* because "we think some people misunderstood the message we were trying to convey. We like the name *The New Democrat* because it leaves no doubt as to our purpose. We're not trying to move the Democratic party to the center, we want to move it forward."[18]

As a 501(c)(4), nonprofit, tax-exempt organization, the DLC was permitted to raise money from any source. In 1990, individual persons could subscribe to the DLC at varying rates from $35 up, and annual subscriptions to the *New Democrat* cost $18. Most of the DLC's funds, however, were donations by private businesses and lobbyists.[19]

The DLC's Mission and Policies

Former Policy Director Bruce Reed described the DLC's essential mission as follows:

> One thing you should understand about the DLC is that despite the fact that it was first founded by southern conservatives, its purpose was not to move the party to the right, although some in the DLC would like that to happen.

The DLC exists to revitalize the Democratic party, and it will do whatever it takes to make that happen. We're not trying to move the party to the left or right, we're trying to move it forward.[20]

The DLC/PPI therefore saw itself primarily as a source of policy innovation for the Democratic party. Its leaders argued that the standard New Deal policies of governmental interventionism and the legacy of 1960s cultural liberalism at home and abroad had proved to be disastrous for the party in presidential politics. Like their Republican counterparts, the DLC has adopted a "progressive" label and in its journals and policy proposals tends to place more emphasis on "new ideas" for moving the party "forward" rather than directly challenging the traditional tenets of Democratic liberalism.

The council's 1990 New Orleans declaration provided the fullest statement that the DLC had yet made of its mission and policy goals:

It declares our beliefs that the fundamental mission of the Democratic party is to expand opportunity, not government; in the politics of inclusion; that America must remain energetically engaged in the struggle for freedom in the world; in deterring crime and punishing criminals; in the protection of civil rights; in the moral and cultural values most Americans share—liberty of conscience, individual responsibility, tolerance, work, faith, and family; and that American citizenship entails responsibilities as well as rights.[21]

The ideas in the New Orleans declaration include "fighting protectionism both in our market and abroad"; "employee stock ownership"; "more workplace democracy"; giving parents "more choice in the schools their kids attend"; "voluntary national service"; a national system of "youth apprenticeship"; "a strategic environmental initiative"; a "guaranteed working wage"; "individual development accounts" to "help low-income families save and to build financial assets"; an "Emerging Democracy Initiative" to "export democratic capitalism and democratic values"; and "restoring progressivity to the tax code." Underlying all these proposals is the theme of the Democrats as the party of "opportunity rather than government."[22]

Even though the original purpose of the council's founders may have been to move the party in a more conservative or centrist direction, by the early 1990s the DLC had moved closer to the neoliberal faction of the party in terms of policy and the ideas earlier associated with Senators

Gary Hart, Paul Tsongas, and Bill Bradley. Also interesting was its lack of emphasis on cultural issues such as abortion, school prayer, and affirmative action, on which the DLC itself was divided and on which a definitive stand by the DLC might deeply antagonize fellow Democrats. Beyond generalities about promoting democracy, the DLC was also deliberately vague on specific foreign policy questions owing to a lack of consensus among its leaders. During the Gulf War debates in January 1991, DLC enthusiasts in the Congress were divided on the resolution authorizing the use of military force with Senators Gore and Robb supporting the resolution and Senator Nunn and Congressman Gephardt opposing it.

At their 1991 annual convention in Cleveland, the DLC passed some more explicit policy resolutions on cutting taxes for middle-income and working families (it endorsed New York Senator Daniel Patrick Moynihan's proposed cut in the payroll tax for Social Security) while raising those on upper incomes, and it renounced quotas and any other form of guaranteed equality while maintaining a commitment to civil rights and affirmative action.[23] The conference and the DLC drew generally favorable notices from the establishment press. According to David Broder of the *Washington Post*:

> For an outfit that was born out of pique, fear and frustration, the Democratic Leadership council has turned into a surprisingly healthy and intelligent 6-year-old. It is contributing positive and practical policy ideas to a party in desperate need of them. . . . The resolutions passed in Cleveland formed a respectable critique of the Reagan–Bush philosophy—and a reasonably clear break from past Democratic thinking.[24]

Broder's praise was echoed in the *Economist*:

> The Progressive Policy Institute, the DLC's in-house think-tank, now produces some of the best work on social policy in Washington, and the resolutions at Cleveland reflected that good sense. They endorsed school choice, free trade with Mexico, and the overarching need for economic growth—all bell-wether themes on which the liberal wing of the party and its allies in the trade unions may find themselves on the wrong side of history.[25]

In this respect one can see a clear parallel between the DLC and the Ripon Society of liberal Republicans. In its heyday in the late 1960s the

Ripon Society also produced innovative policy proposals that were widely praised by the national news media. Many of these proposals, like a volunteer army, welfare reform, and revenue sharing, were actually adopted by the Nixon administration, but all of this had little relevance to the Ripon Society's long-term goal of reestablishing moderate or progressive Republicanism as a national political force.[26]

The DLC was able to gain plaudits from the national press corps and Washington insiders for its "innovative" policy proposals, but without the national cadre of grass-roots activists that could be mobilized behind a presidential candidate sympathetic to the aspirations of the DLC, all those policy proposals and impressive analyses would not change the overall tone and direction of the national party. The contests of the 1980s demonstrated that presidential elections are won and lost not on the fine print of the party platforms but on one or two themes or a rhetorical tone that defines the parties in the minds of the voting public at that particular time. The DLC might be able to place worthy planks in the party platform, but would it be able to change the overall tone and direction of the party, something that the Ripon Society had singularly failed to achieve inside the GOP?

The DLC at the Outset of the 1992 Campaign

Entering the 1992 election cycle there was little question that the DLC had made some impact on the national political scene. It had been a major source of innovation for the Democrats in domestic policy, and it had attracted some of the most able and attractive Democratic officeholders into its ranks. In short, the DLC served as a forum and a rallying point for moderate-to-conservative Democrats in an era when they appeared to have been marginalized within the national party.

Nevertheless, the DLC had probably reached the limits of its effectiveness in national politics unless it could move from the realm of policy formation and research in Washington, DC, to organization and strategy on a national level. In 1989 the PPI published *The Politics of Evasion*, by William Galston and Elaine Ciulla Kamarck, which provided an excellent analysis and summary of the reasons for the Democratic debacle in the presidential elections of the 1980s, namely, the desertion of the white middle and working classes to the GOP.[27] The problem had been in moving from analysis to prescription. If the Democratic party's nominees were regularly driving elements of the party's core constituency out of the Democratic ranks, there had to be something wrong with the process that selected those nominees. It was the structural advantages enjoyed by the single-issue groups and the ideological activists—the "liberal fundamentalists" in the DLC's terms—that produced the ineffectual presidential candidates who were unrepresentative of the "mainstream" of the party at the mass level. The next logical stage in the DLC's development, therefore, should have been a radical restructuring of the nominating process so as to give its favorites the opportunity to secure the presidential nomination. Although they seldom publicly acknowledged this, for fear of offending key constituencies in the party elite, the DLC leadership was not unaware of the problem.

When challenged as to why the DLC had not pursued radical reform of the party rules, Al From replied:

> If we wait for that, it will never happen. We think that until the party understands why it gets beat in presidential elections, they're not going to do anything about it. We want to give the party a kind of realism therapy. In this election cycle for the first time, most Democrats have a better understanding of that . . . the DNC [Democratic National Committee] is a disaster, but they happen to have control over the rules of the nominating process, and they play the very kind of politics that brought us down to landslide defeats. . . .
>
> Having said that I don't believe that the rules completely determine the outcome. In 1984 the rules were written for Mondale, but Hart would have won if he hadn't been such a flake.[28]

From did acknowledge, however, that the DLC's creating state chapters was intended to build a national infrastructure that a moderate-to-conservative Democratic presidential candidate could use:

> Our goal is to have chapters in twenty-five states within a year. We hope that this will be a vehicle through which we can get people to become delegates. It should also provide a forum where mainstream candidates can comfortably go and promote their message, and over time we can change what the party stands for and create a climate for a presidential candidate in 1992 or 1996. In essence the strategy we're pursuing is to develop a message that a candidate can run on and an infrastructure that a candidate can pick up.[29]

Because the DLC's process of developing a national infrastructure was still in its early stages as the 1980s drew to a close, it continued to be very much an elite-oriented, Washington-based organization. In 1990 the DLC counted 10 of the 28 Democratic governors (36 percent), 28 of the 55 Democratic senators (51 percent), and 112 of the 260 Democratic House members (43 percent) among its members.[30] Among the party's governing elite—its elected officeholders—the DLC was strong, but the party's presidential elite is an entirely different grouping of interest-group leaders, fund-raisers, and activists, and here the DLC's strength was not nearly so evident. The council's Washington orientation was rather unfortunately reinforced by the press accounts of its 1991 convention in Cleveland, all of which emphasized the high number of corporate Washington lobbyists among the voting delegates. Even the generally sympathetic David Broder conceded:

> The Cleveland convention did not look or feel like a Democratic gathering. You looked around the floor and saw few teachers or union members, few blacks and fewer Hispanics.
> In their place you had dozens of corporate lobbyists who pay the DLC's bills in return for access to its influential congressional members and governors. Many of the lobbyist-delegates acknowledge being Republicans; one was a top staffer for Spiro Agnew. But they were in there voting on resolutions, just as if they really cared about the Democrats' winning.[31]

The DLC's apparent weakness among the party's key constituent groups in the presidential nominating process—its critics in the party frequently referred to the DLC as the "white boys' caucus" or the "white men in suits"—was obviously problematic in regard to establishing a national base of support.

The DLC faced yet another problem when its liberal fundamentalist adversaries in the Democratic party began to countermobilize. In February 1991 some five hundred Democrats formed the Coalition for Democratic Values (CDV) led by Rev. Jesse Jackson and Senators Howard Metzenbaum (Ohio) and Tom Harkin (Iowa) in an effort to counter the DLC and reemphasize the party's commitment to economic populism, as opposed to "shadow Republicanism."[32] During the DLC's Cleveland gathering, Jesse Jackson—who was not invited to the DLC meeting—held his own meeting in a union hall, as if to contrast his following with the white-collar group at the DLC. Democratic National Chairman Ron Brown, who did address the DLC gathering, used the opportunity to chastise the DLC for promoting factionalism within the party.[33]

One of the most serious and persistent critics of the DLC and its role within the Democratic party was the journalist Robert Kuttner. Kuttner and the renegade Republican commentator Kevin Phillips were the leading advocates of a new politics of economic populism as the answer to the Democrats' problems.[34] They argued that Democrats should cease their neoliberal flirtation with Republican economics and come home to the social democratic policies of governmental interventionism that had served them so well during the New Deal era. According to Kuttner, the policy proposals of the DLC and the PPI were not substantially different from those of their liberal fundamentalist rivals, and therefore, the sole *raison d'être* of these organizations was to refight old factional battles within the Democratic party:

> It's time for a genuine truce. Republicans learned to stop beating-up on each other a long time ago. The DLC attacks liberals for fundamentalist litmus tests. But the liberals are in a newly pragmatic mood, and the DLC are the ones using party rules, minimum wage, Jackson-bashing, and memories of ancient faction fights as litmuses.[35]

The Kuttner/Phillips analysis was correct up to a point: Economic populism remained the source of the Democratic party's popular appeal. Large segments of the American electorate continued to support the social programs of the New Deal–Great Society period, and they even were prepared to see the expansion of such programs in areas such as education and health care. The problem that Kuttner and Phillips ignored was that the Democrats' identification with a particular set of cultural issues and symbols in presidential politics after 1968 is what had driven the "Nixon, Reagan, and Bush Democrats" to the GOP.[36] In this respect, Al From's rejoinder to the Kuttner/Phillips analysis was correct:

> There's a degree to which economic populism can be effective, but statism is never going to sell in this country. We're never going to have a Swedish planning system. . . .
>
> The other answer to Kuttner is that he greatly underestimates the social issues and cultural issues. I believe in the [Michael] Barone theory. Except in economically very bad times, the cultural divide is greater in American politics than the economic divide.[37]

Although From's analysis is more accurate than Kuttner and Phillips's, it was by no means clear that the DLC was following his prescriptions.

The DLC's leadership had striven mightily to avoid a conservative label in favor of a progressive one, and the DLC's policies tended to reflect a blend of neoliberal and populist economics that was not greatly at odds with the proposals of the liberal fundamentalists, as Kuttner and Phillips had observed. Beyond opposing racial quotas, the DLC had chosen to ignore social and cultural issues as far as possible in the hope that with a recession and a more reassuring candidate at the top of the ticket, those issues would become irrelevant (and in 1992 they were proved to be correct!).

Unfortunately, the Republican party had become very accomplished at forcing Democratic presidential candidates to stand up and be counted on issues like racial preferences, the death penalty, and "patriotism." In 1988 the Dukakis campaign had tried to suppress the cultural issues and to discuss neoliberal economic ideas, but the strategy was a disastrous failure.

On the other hand (as the 1992 Clinton campaign realized), for the DLC to adopt a more explicitly conservative tone on social and cultural issues would have reduced its support by women and minority Democratic officeholders and would have confined its ranks to a small cadre of largely southern, white, conservative males — not a large enough group to play a significant part in the national party.

In truth, the difference between the DLC and its liberal fundamentalist opponents appeared increasingly to be one of tone and emphasis rather than real substance. And here is yet another example of the ideologization of American party politics over the past quarter-century or so. Whereas the conservative Democrats of the 1940–60 period were more conservative than most of the Republicans on many issues, a glance at the *Mainstream/New Democrat* showed that the ideological chasm between the DLC and the Republican party of the late 1980s was vast. In such circumstances, it is not clear that even with a DLC stalwart like Charles Robb, Sam Nunn, or Dick Gephardt at the head of the Democratic ticket in 1988, the outcome would have been different.

Conclusion: The Absent Center

The core of the DLC's problem before the 1992 elections was that the Democratic party's governing elite, where it was strong, appeared to have little influence over the party's presidential elite, where the DLC appeared to be weak. Among the party's upper echelon of elected officeholders at the local, state, and federal levels, the DLC had been influential, but officeholders have very little influence over the presidential nominating process, and the DLC's essential task was to elect a Democratic president. The DLC's establishment of state chapters was a move in the direction of the national grass-roots infrastructure that its Republican counterparts in the Ripon Society were never able to build.

Building such an infrastructure looked likely to be an extremely formidable task, however, since post-1968 presidential nominating politics placed a premium on mobilizing interest groups, fund-raisers, and activists, none of which responds enthusiastically to "moderate," "technocratic," "pragmatic" candidates with high reputations inside the capital beltway. In fact, moderation or pragmatism has proved to be a rather poor means of both political mobilization and governance in modern American politics. The contrast between the administrations of Ronald Reagan and Jimmy Carter is instructive in this regard. Reagan's greater political success can largely be attributed to the fact that the rhetorical and ideological tone of his administration was a much more effective rallying point in terms of congressional and public support than was the technocratic style of the Carter White House.

When Daniel Bell proclaimed the "end of ideology" in the late 1950s, he was correct as far as western Democracies in general were concerned, but wrong with regard to American party politics, in regard to which the post-1964 period has been labeled the "age of ideology."[38] Seymour Martin Lipset noted that whereas the Social Democratic parties of Western Europe had been moving consistently rightward over the past two decades, the tendency in the United States was in the opposite direction:

But if the democratic socialist stories in Europe and Asia have been of left recognition of the need for market-oriented economic policies to stimulate investment and productivity and of electoral efforts to take the middle of the road voters away from the right, the American one has been curiously different. . . . In recent years, the national Democrats have been more disposed to adhere to a redistributionist, progressive tax, antibusiness orientation than many, if not most, social democratic parties. They have also been more supportive of cultural liberalism, of practices invidiously dubbed as "permissive."[39]

Although I am dissatisfied with Lipset's implication that the Democrats have become more "left" in economics (the evidence seems to be in the other direction, perhaps to the detriment of the party's electoral appeal), he certainly is correct with regard to foreign policy and cultural issues.[40] Lipset's explanation for this "American exceptionalism" also gets to the heart of the matter: The parliamentary governmental systems of the other major industrial democracies place a premium on strong, hierarchical, and disciplined party organizations, something that the United States, with its separation between the executive and legislative branches, has always conspicuously lacked.[41] The problem became even more evident after 1968, with the disintegration of the old party machines and the triumph of the "new politics." There was no center or centers of power in the Democratic (or Republican) party that could effectively enforce a change in ideological direction, and thus candidates in both parties — particularly in presidential races — were driven toward the network of more militant single-interest and ideological groups that provided the money and organization for presidential campaigns. In such an atmosphere, Washington-oriented, technocratic, organizations like the Ripon Society and the DLC found it much harder to prevail.

With the rise of ideological politics, the Democratic party lost "middle America" — the lower-middle and working-class, white, northern ethnics and white southerners — who perceived the presidential Democratic party as focusing on the oppression of racial minorities, women, and the Third World by the white male power structure but as seeming to have little interest in their fundamental concerns — income, family, neighborhood, and crime.[42] The phenomenon of an almost permanently divided government, or "split-level realignment," in national politics arose during the 1968–92 period because these voters largely stayed with congressional, state, and local Democratic candidates who purveyed the old New Deal liberalism of providing popular social programs and bringing home projects to their people. At this level the Republican emphasis on cost cutting, minimal government intervention, and free-market solutions to social problems had a very limited appeal.[43]

Thus the neoliberal economics that featured so prominently in the DLC's list of policy prescriptions seemed unlikely to be a certain vote winner for the Democrats, because the swing group of voters still generally approved of traditional Democratic attitudes in this sphere. To win them back, the party needed at least to tone down the cultural and foreign policy liberalism it had espoused since the mid-1960s and emphasize its economic populism, but it could not do one without the other, as Kuttner, Phillips, and the DLC strategists appear to believe. (The Clinton campaign surely learned this lesson in 1992.)

In contrast with fifty years ago, when consensus was the essential criterion for agreeing on a presidential nominee, the middle of the road was almost the worst position for presidential aspirants after 1968. The weakness of the center in national politics, of course, may have had negative consequences for the political system as a whole.[44] It certainly polarized society on issues such as race and sexual conduct and encouraged a divided government in Washington (because the electorate was reluctant to trust either of the two ideologized national parties with total control over the federal government) that in turn has led to a lack of accountability to the electorate and a permanent deadlock on crucial policy questions. The situation can probably be permanently changed only by radically restructuring the nominating process so as to make it more responsive to those Democratic and Republicans who appeared to be more or less excluded from it at both the elite and the mass levels between the 1968 and 1988. Restoring the national conventions to their former role and abolishing the primary system altogether might have been the best solution but, given the current social and political context, is likely to be a pipe dream. A more plausible solution might be to move in the opposite direction, toward a national primary and runoff, which might place the more experienced and representative candidates at an advantage, although the logistical and constitutional difficulties in establishing such a process appeared to be insurmountable as the 1990s dawned.

6

The 1992 Election: The South Recaptures the Democratic Party (and the White House)

The 1992 election marked a crossroads in American politics in more than one respect. With the end of the cold war and the collapse of Communism, the Republican presidential advantage in foreign defense policy was nullified. A sluggish national economy for most of the Bush presidency also put economic pressure on the middle- and working-class voters who had deserted the Democrats on foreign policy and cultural issues during the 1980s. These fears combined with a widespread national anxiety over America's relative economic "decline" vis-á-vis Japan and Western Europe.[1] Finally, George Bush, in contrast with his Republican predecessor, appeared to have little empathy with middle- and working-class Americans and was widely perceived to be a weak and vacillating leader.

The Democrats still had to grasp the opportunity, however, and their presidential travails over the previous two decades did not inspire great confidence that they would finally be able to seize the White House in 1992. That they did so can be credited to an unprecedented mobilization of the national party elite behind the candidacy of the southern moderate Arkansas governor, Bill Clinton, whose campaign strategy was brilliantly designed to take maximum advantage of the national anxieties of 1992. Clinton's success was also due to very weak opposition from the New Left and neoliberal sections of the party and a background that enabled him to appeal to the neoliberal and southern conservative Democrats simultaneously. His success revived the southern wing of the Democratic party and brought the DLC and its PPI think tank into major positions of power and authority in Washington, DC.

Nonetheless, Clinton was elected a minority president in a three-way contest, with only 43 percent of the national popular vote, and it was by no means clear that the revival of the southern moderates would be any more permanent than it was after Carter's victory in 1976. Indeed, despite the presence of two southerners on the national Democratic ticket, the South was Clinton's weakest region in November, and he carried only his home state of Arkansas, vice-presidential nominee Albert Gore's Tennessee, Louisiana, and (very narrowly) Georgia. This chapter explains how Clinton won the nomination and the presidency in 1992 and how he was able to overcome the structural bias against "moderate/centrist" candidacies in the contemporary presidential nominating process.

A Big Fish in a Small Pond: The Democratic Field in 1992

Despite the evident weaknesses of President George Bush, the Democratic field of candidates in 1992 was probably one of the weakest in recent years. The main reason for the reluctance of leading Democrats to enter the fray was Bush's triumph in the Gulf War in the spring of 1991 and the astronomical presidential approval ratings that Bush enjoyed in the wake of that conflict.[2] As in 1988, many of the Democratic party's most prominent national figures decided to sit out the 1992 race. The absence of Rev. Jesse Jackson from the contest in 1992 (owing to Bush's apparent strength and Jackson's reluctance to appear as a three-time presidential loser) critically weakened the New Left segment of the party.[3] The nearest surrogates for Jackson were the populist Iowa senator Tom Harkin and the former California governor Jerry Brown. But neither had the appeal to black Democrats that Jackson possessed. Harkin was burdened by the lack of a national identity, close ties to organized labor, and a strongly protectionist position on trade policy.[4] Jerry Brown's quixotic candidacy was not taken seriously by political insiders, who had long ago written him off as a dilettante and political chameleon.[5]

Among the Democratic neoliberals, Senator Bill Bradley of New Jersey was probably the contender with the highest national profile before the 1990 congressional elections, but his surprisingly narrow reelection to the Senate and Bush's triumph in the Gulf War (which

Bradley had opposed) persuaded him that he would do better to consolidate his political base before embarking on a national campaign.[6] In the
spring of 1991 there was considerable speculation that Senator John D.
("Jay") Rockefeller, IV, who had been attracting the attention of the
national media through his campaign for health-care reform, would enter
the presidential race.[7] But the senator subsequently got cold feet and
withdrew from the race in the summer. Senator Robert Kerrey of Nebraska, who did enter the ring in late 1991, appeared to be a potentially
strong candidate. A Vietnam veteran who had later campaigned against
the war, Kerrey seemed like the kind of charismatic "outsider" who
would receive positive media coverage and do well among the quirky
New Hampshire Democratic primary electorate.[8] His liabilities, however, included a lack of experience in national issues, several inconsisten-
cies in his positions on various issues, and doubts about his temperament
with regard to the office of the presidency.[9]

One of the first of the neoliberal Democrats in the 1970s, former
Massachusetts Senator Paul Tsongas, had had his political career trag-
ically interrupted in 1984 when he was diagnosed with lymphatic cancer.
Having staged a remarkable recovery, Tsongas was the first Democrat to
declare for the presidency in 1991, and although he was rapidly dismissed
by media pundits as "another liberal Greek from Massachusetts," it soon
became clear that his campaign—oriented around the need for tough
measures to deal with the budget deficit and strong liberal stands on
social and cultural issues such as abortion and gay rights—possessed a
particularly strong appeal to the neoliberal constituency of suburban
professional Democrats.[10] Moreover, the presidential primary season
would begin on Tsongas's home turf of New England.

Perhaps the most prominent of all the Democrats on the neoliberal/
New Left side of the party was New York Governor Mario Cuomo. For
months Cuomo kept the media and his fellow Democrats guessing about
his intentions for 1992, and his hesitation probably served to keep some
other Democrats such as Bradley and Rockefeller out of the race.[11]
Cuomo finally ended his agonizing over the presidency by announcing
that he would not be a candidate, on the very day of the filing deadline for
the New Hampshire primary.[12]

The absence of Cuomo, Bradley, and Jesse Jackson left a void of
leadership and a field of largely unknown and untested candidates on the
left/liberal spectrum of the Democratic party. It was thus apparent that if
the southern section of the party could rally its forces behind a single

candidate, who could also appeal to the dwindling but still substantial number of old-fashioned regular Democrats (who had no obvious champion to support) and the more pragmatic neoliberals, they might be able to turn this situation to advantage. Yet as in 1988, many of the South's "top guns" failed to make the race. Despite his strength as the vice-presidential candidate in 1988, Texas Senator Lloyd Bentsen was generally thought to be too elderly at seventy-one and too closely linked to the corporate world to be a possible national nominee for the Democrats. Georgia Senator Sam Nunn had voted against the Gulf War—thus sacrificing some of his credibility to conservative voters—while still remaining too far from the Democratic mainstream on other issues to be a plausible presidential candidate.[13] Senator Charles Robb of Virginia had supported the Gulf War, but his presidential aspirations had been set back by tawdry scandals emanating from a bitter feud with Virginia's Democratic governor, L. Douglas Wilder.[14] Missouri Congressman Richard Gephardt had also opposed the war and, given Bush's strong standing, was more eager to concentrate on developing his career in the U.S. House, where he had risen to the second-ranking position of majority leader.[15] By far the most interesting possible southern contender was the aforementioned Governor Douglas Wilder of Virginia, the nation's first elected black state governor and a fiscally conservative and socially moderate Democrat who might be able to add white conservative support to a potentially heavy black voting base. But personal difficulties and disorganization led to Wilder's rapid withdrawal from the race even before New Hampshire.[16] Finally, Senator Albert Gore of Tennessee had supported Bush on the Gulf War and appeared to have strengthened his position with southern conservatives, but Gore also was reluctant to risk another failed nomination campaign that might end his presidential prospects for good.[17]

The one remaining plausible southern moderate for conservative Democrats and the DLC to rally around was the DLC's chairman, Governor Bill Clinton of Arkansas. Clinton's credentials as a neoliberal were perhaps even stronger than his credentials as a southern moderate. A Rhodes scholar educated at Georgetown and Yale Law School, who had protested against the Vietnam War and managed George McGovern's campaign in Texas in 1972, Clinton returned to Arkansas and won election as governor of the state in 1978 at the tender age of thirty-two. After an unexpected defeat in 1980, Clinton came back and won reelection as governor in 1982 and held onto the office for the next ten years. As

Arkansas governor, Clinton epitomized the "New South" reformist generation in a state that had almost the worst record in the region in regard
to poverty and poor service.[18] His identification with a more "pragmatic"
approach to governing and his shedding of his and his wife Hillary's
cultural liberal associations (which had apparently contributed to his
1980 defeat) endeared Clinton to traditional white southerners as well as
neoliberal reformers.[19] With Jackson and Wilder out of the race, Clinton
(for the same paradoxical reasons as his fellow white Southern Baptist
Jimmy Carter in 1976) also had potentially wide appeal to black Democrats nationwide.[20] Despite Arkansas's remoteness from the national
political scene, Clinton had already established a national political profile as an appealing and capable, southern, moderate reformer.[21]

    Clinton had also done his cause considerable good through his service
as chair of the DLC in 1989–91. This gave him a national platform and a
national network of support for a presidential campaign (as mentioned in
Chapter 5, Clinton had pushed hard to create more than twenty state
chapters of the DLC during his time as chairman). However, we should
reemphasize that Clinton's political profile was also calculated to appeal
to both neoliberals and regular Democrats. On civil rights and social
issues like abortion, Clinton took clearly liberal positions while simultaneously stressing an appeal to economically pressed "middle-class
families." On foreign policy Clinton had not been forced to take positions
as Arkansas governor, whereas his opponents all had taken strong anti-
interventionist positions on most foreign policy issues, particularly the
Gulf War. No other contender was thus so uniquely placed as Clinton was
to straddle the factional divisions and appeal to all segments of the
Democratic party.[22]

    In contrast with 1988, conservative southern DLC forces had to rally
behind Clinton because there was no alternative in the field. This meant
that if the Clinton campaign could just survive with a decent showing in
the difficult terrain of New Hampshire, Clinton could earn a potentially
enormous windfall of Democrats in the southern primaries that were still
largely concentrated in a two-week period in early March (see Table
6.1).[23] The absence of Jackson and Wilder made the potential windfall
even greater. Finally, to direct his campaign strategy, Clinton hired the
two political consultants, James Carville and Paul E. Begala, who had
engineered the upset victory of Senator Harris Wofford over former
Attorney General Richard Thornburgh in the November 1991 special
Senate election in Pennsylvania, a race that had been turned around by the

Table 6.1. The Democratic Primary and Caucus Schedule, 10 February to 10 March 1992 (Super Tuesday)

Date	State	Percentage of National Delegate Total	Cumulative Percentage
2/10/92	Iowa (c)[a]	1.3	1.3
2/18/92	New Hampshire	0.6	1.9
2/23/92	Maine (c)	0.7	2.6
2/25/92	South Dakota	0.5	3.1
3/3/92	Colorado	1.3	4.4
3/3/92	Georgia	2.1	6.5
3/3/92	Idaho (c)	0.6	7.1
3/3/92	Maryland	1.9	9.0
3/3/92	Minnesota (c)	2.0	11.0
3/3/92	Utah (c)	0.7	11.7
3/3/92	Washington (c)	1.9	13.6
3/5/92	North Dakota (c)	0.5	14.1
3/7/92	Arizona (c)	1.1	15.2
3/7/92	South Carolina	1.2	16.4
3/7/92	Wyoming (c)	0.4	16.8
3/8/92	Nevada	0.5	17.3
3/10/92	Delaware (c)	0.4	17.7
3/10/92	Florida	3.7	21.4
3/10/92	Hawaii (c)	0.6	22.0
3/10/92	Louisiana	1.6	23.6
3/10/92	Massachusetts	2.5	26.1
3/10/92	Mississippi	1.1	27.2
3/10/92	Missouri (c)	2.0	29.2
3/10/92	Oklahoma	1.2	30.4
3/10/92	Rhode Island	0.7	31.1
3/10/92	Tennessee	1.8	32.9
3/10/92	Texas	5.0	37.9

[a] (c) caucus state

Source: Rhodes Cook, *Race for the Presidency: Winning the 1992 Nomination* (Washington, DC: Congressional Quarterly Press, 1991).

Wofford campaign's emphasis on the issues of economic recession and health care and that also turned out to be a harbinger of the 1992 presidential general election campaign.[24]

Having been established by the media as the early front-runner in the fall of 1991, Clinton's problem was to survive the prolonged and detailed scrutiny of his political career and private life that seems to be endemic to modern American presidential politics.[25] He would also have to survive the concentrated assaults of his fellow candidates and the perception that no Democrat could possibly defeat a Republican presidential incumbent who had just won a foreign war.

The Democratic Primary Campaign

Perhaps surprisingly, the most serious challenge to Clinton in the early primary season came from Paul Tsongas. As mentioned earlier, Tsongas's combination of fiscal conservatism and social and foreign policy liberalism found a strong resonance among the neoliberal, professional, suburban Democratic constituency that had persistently responded to "new politics" Democrats such as McCarthy, McGovern, Carter, Hart, and Dukakis since 1968.[26] By the time of the New Hampshire primary, it was clear that none of the other contenders had caught fire. Harkin appeared too shrill and too closely associated with old-time New Deal liberalism; Kerrey's campaign lacked a clear definition and theme; and Jerry Brown's low-budget effort was not taken seriously.[27] The Democratic party establishment—the leading officeholders and leaders of powerful interest groups—was already beginning to coalesce quietly around Clinton in the fall of 1991, and this tendency became more pronounced as it became clear that his strongest rival was Tsongas, whose narrow geographical base in New England, economic conservatism, and ethnic and geographic associations with the disastrous Dukakis ticket in 1988 made him especially unappealing to Democratic regulars, southerners, and New Leftists.[28] If any Democrat's chances in November were slim, then those for a Democrat with Tsongas's pedigree seemed particularly so.

In the fall of 1991 Clinton had appeared to be comfortably placed. He rolled through a series of straw polls at various state Democratic party gatherings, and early opinion soundings from New Hampshire showed him to be in the lead.[29] The Clinton campaign faced its hardest test when the news media aired the allegations of Gennifer Flowers, a former nightclub artiste and Arkansas state employee, that the Arkansas governor had conducted a long-standing extramarital affair with her. The Clinton campaign reacted by counterattacking the news media for airing such a sordid and irrelevant story when the voters wished to debate the serious issues of economic recession and competitiveness facing the country and also by having the candidate and his wife talk frankly about their marriage on national television.[30] No sooner had the Flowers story been surmounted, however, when the Clinton campaign faced a further barrage of negative publicity concerning Clinton's avoidance of the military draft as a Rhodes scholar in Oxford, England, at the tail end of the Vietnam War.[31]

These stories dominated the New Hampshire campaign as Tsongas overtook Clinton in the New Hampshire opinion polls.[32] What saved Clinton was the inability of any of the other contenders in New Hampshire to capitalize fully on his weakened position and grab second place (it was by now taken for granted that Tsongas would win). The liabilities of Harkin, Kerrey, and Brown remained so great that even a greatly weakened Clinton was preferable to those large sections of the party that could not stomach Tsongas. The eventual result of the primary (see Table 6.2), with Clinton finishing only eight points behind Tsongas (rather than the twenty-point spread that had been feared at one time), was shrewdly converted by the Clinton campaign into a triumph of unexpected dimensions for a candidate who had been on the ropes two weeks previously. Endless news footage of Clinton as New Hampshire's "comeback kid" gave him the momentum he needed for the crucial primary tests on his home turf in the South.

Poor showings in New Hampshire effectively killed off Kerrey and Harkin, and the battle had essentially been reduced to a two-man contest between Clinton and Tsongas (Jerry Brown's candidacy was still generally regarded as little more than an irritation), both essentially "new politics" neoliberals with a different regional focus. The dynamics of the campaign

Table 6.2. Results of the New Hampshire Democratic Primary, 18 February 1992

Candidate	Percent
Tsongas	33
Clinton	25
Kerrey	11
Harkin	10
Brown	8
Others	13
	100

Source: Official returns, reprinted in *Congressional Quarterly Weekly Report*, 4 July 1992, p. 69.

had, however, altered Clinton's strategic position. When Clinton originally announced, he had envisaged running as the "neoliberal" alternative to a northeastern regular like Mario Cuomo. But with Cuomo's nonappearance and the emergence of Tsongas as the favorite of the neoliberal, suburban, professional constituency, Clinton was compelled to run as the regulars' candidate, with support from southerners (for cultural reasons) and New Leftists (for ideological reasons) if it came down to a straight contest between the two.[33] With the southern primaries approaching, this apparently placed Clinton in an impregnable position.

With the experience of 1988 in mind, however, Tsongas had several advantages in the primary schedule. Before Super Tuesday there were primaries and caucuses in several northern and western states where Tsongas might be expected to run strongly, such as Maine, Maryland, Colorado, Washington, Massachusetts, and Rhode Island (see Table 6.1).[34] The 1988 reelection campaign had also demonstrated that the clutch of Super Tuesday primaries in the South would not necessarily deliver the desired bloc vote to the regional favorite. There were plenty of places in the South where Dukakis had run strongly in 1988—suburban and migrant-dominated Florida, high-tech North Carolina, the Atlanta suburbs, and the college towns—where Tsongas also had the potential to do well.

Clinton thus still had to work hard to deliver his home region, but he nevertheless had several advantages denied to Gore and Gephardt in 1988. For a start, the absence of Jesse Jackson, who had corralled most of the southern black vote in 1988, gave Clinton a tremendous advantage. Although the black turnout might be reduced, those blacks who did vote would vote overwhelmingly for Clinton, who had a strong regional and cultural appeal to black voters. Moreover, Tsongas's campaign, with its strong emphasis on fiscal rectitude and economic retrenchment, had little appeal to minority voters, many of whom were heavily dependant economically on federal social programs.[35] For similar reasons, the largely Jewish elderly Democratic primary electorate in southern Florida, which had been attracted to the northeastern liberal Dukakis in 1988, were uneasy that Paul Tsongas's harsh economic medicine might include reductions in their social security benefits, an anxiety that was fully exploited by the Clinton campaign in this crucial state.[36] Finally, the moving of the Georgia and South Carolina primaries to 3 and 7 March, respectively, enabled Clinton to translate his comfortable victories in both states (see Table 6.3) into a major degree of momentum for the Super Tuesday contests on the following Tuesday (10 March).

Table 6.3. Results of Junior Tuesday Primaries, 3 and 7 March 1992 (percent)

Candidate	Colorado	Georgia	Maryland	South Carolina
Brown	29	8	8	6
Clinton	27	57	34	63
Harkin	2	2	6	7
Kerrey	12	5	5	1
Tsongas	26	24	41	18
Others	2	–	1	3
Uncommitted	2	4	6	3

Source: Official returns, reprinted in *Congressional Quarterly Weekly Report*, 4 July 1992, p. 69.

Clinton was also lucky that Tsongas was unable to exploit his regional and cultural base as effectively as he might have done, owing to the surprising resilience of Jerry Brown's candidacy. In states such as Maine and Colorado, which should have been prime territory for Tsongas, Brown's shoestring "antiestablishment" campaign had a strong appeal to countercultural/New Leftist voters who formed a significant portion of the active Democrats in such states.[37] By siphoning off such support from Tsongas, Brown weakened the former senator's position vis-à-vis Clinton. In the desperate scramble for votes across a vast expanse of territory, Clinton's formidable campaign infrastructure also paid off for him. The Clinton campaign already had money, television advertising, and state and local officeholders in place in the southern states before the New Hampshire primary. Tsongas, who had concentrated all of his resources in New Hampshire, had nothing like the resources that Clinton had for Super Tuesday.[38]

In what turned out to be an uneven contest, Clinton made a clean sweep in the Super Tuesday primaries in the South, winning huge victories across the region (see Table 6.4). This propelled him into a formidable lead in delegates over those won by Tsongas and Brown. The next big tests in Illinois and Michigan were critical. If Clinton could demonstrate strength outside his home region and win those two states, then his path to the nomination looked secure. This midwestern, "rust-belt" territory, containing many unionized Democrats in traditional manufacturing industries who were particularly concerned about competition from Japanese imports, was not particularly receptive to Paul Tsongas's austere, militantly pro-free-trade economics.[39] Although Jerry Brown secured significant local union support by eagerly coming down for more managed trade and a rejection of the North American Free Trade Agreement

Table 6.4. Overall Outcome of "Super Tuesday" Primaries, 10 March 1992

Candidate	Primaries Won	Total Vote (%)	Delegates Won
Clinton	8	54	432
Tsongas	2	28	210
Brown	–	11	25
Others and uncommitted	–	7	33

Source: *Congressional Quarterly Weekly Report*, 14 March 1992, p. 632.

(NAFTA), Clinton, not wishing to alienate the pro-free-trade Tsongas suburban neoliberal constituency, took a more ambivalent position.[40] Electorally this proved to be the wisest course, as he easily defeated Tsongas in Illinois and Brown in Michigan. Tsongas reacted by announcing his withdrawal from what appeared to be a hopelessly uneven contest, thus effectively ensuring Clinton's nomination.[41]

There still were several obstacles for the Clinton campaign to surmount before the prize was firmly nailed down. The Flowers and draft allegations still had a negative effect on Clinton's national ratings with Republican and Independent voters, and fears of further revelations and the Republican attack machinery led many senior Democrats to doubt whether Clinton, despite an increasingly favorable press and rave notices from Washington insiders as the most formidable Democratic campaigner since John F. Kennedy, could run a serious race against George Bush in the fall.[42] Rumors continued to circulate in the press about the possible late entry into the race of a senior national Democrat such as the Texas senator and 1988 vice-presidential nominee Lloyd Bentsen, Congressman Richard Gephardt, or even (yet again) Mario Cuomo, should the Clinton campaign collapse.[43]

The unlikely vehicle on which such speculations rested was the ever-enigmatic and surprisingly resilient Governor Jerry Brown, the former neoliberal hero of the 1970s now running for president on an unabashedly New Left platform.[44] If Clinton should lose a Democratic primary election in a major state to a candidate with as many liabilities as Brown, then legitimate questions would be raised about Clinton's viability against Bush in November. Clinton grew more nervous when Brown secured a surprise victory in the Connecticut primary, the dress rehearsal for New York. In the dynamics of the 1992 campaign, New York looked like possibly fertile territory for Brown. He had a potential base of support among New York City New Left Democrats, union leaders, and minority activists, and white Southern Baptist presidential candidates like Clinton had traditionally been regarded with some suspicion by New York's cosmopolitan and polyglot electorate (Carter lost the primary in New York in both 1976 and 1980).

Once again, however, the Clinton campaign showed itself to be most formidable when pushed into a tight corner. Clinton vigorously counterattacked Brown in televised debates for raising the issue of Hillary Clinton's business associations in Arkansas, and Clinton also attacked Brown's "flat-tax" proposal as likely to damage middle- and working-class voters.[45] Brown's courting of the minority vote by offering Rev. Jesse Jackson the vice-presidential nomination gained him little encouragement from Jackson but guaranteed a high level of hostility from the large Jewish primary electorate in New York City.[46] Brown's concentration on New York also lessened his chances in the even more fertile New Left territory of Wisconsin and Minnesota, which held primaries on the same day.[47] Again, the formidable national organization that enabled Clinton to compete effectively in all three states turned out to be a tremendous asset. Finally, Brown's chances of defeating Clinton in New York were probably fatally damaged by a flurry of speculation that Paul Tsongas (whose name was still on the New York ballot), would reenter the race in the event of a Clinton defeat.[48]

Clinton's convincing victory in the Empire State (with noncandidate Tsongas edging Brown into third place) effectively ended the Democratic primary campaign. By demonstrating strength in major states outside his southern base, Clinton finally snuffed out potential trouble from more established Democrats. The effectiveness of the Clinton campaign in its ability to surmount formidable challenges and crises had also been impressive. After New York, Clinton had such a formidable lead in the delegate count that lacking some catastrophic revelation, he was the certain Democratic presidential nominee. Brown stayed in the race and continued to harry Clinton through the rest of the primaries, but with no serious prospect remaining of impeding Clinton's progress toward the nomination, the Californian was little more than a nuisance candidate (see Table 6.5).

The question that remains is why Clinton—the self-described southern moderate and DLC champion—was able to buck the trend of recent presidential nominating politics (set out in Chapter 1) and win the Democratic presidential nomination from a centrist position. Moreover,

Table 6.5. 1992 Democratic Primary Vote by Time Period (percent)

Candidate	New Hampshire to to 3 March	Super Tuesday to Illinois (3/10–3/17)	Connecticut to California (3/24–6/9)	Total
Clinton	38	53	53	52
Brown	11	13	25	20
Tsongas	31	27	11	18
Others and uncommitted	20	7	11	10

Source: Congressional Quarterly Weekly Report, 4 July 1992, p. 72.

did Clinton's nomination portend a fundamental change in the structure of presidential nominating politics that would favor more moderate candidates in both parties during the 1990s, and did it change the essential dynamics of the presidential nominating process?

The answer to the second question appears to be negative. Clinton won in 1992 for reasons that were specific to the circumstances of that campaign and his own candidacy. The most obvious reason for his success was the weakness of the Democratic field in 1992, particularly on the left side of the party. Better-known New Left or neoliberal candidates such as Jesse Jackson, Mario Cuomo, George Mitchell, Patricia Schroeder, and Bill Bradley, around whom those segments of the party might have rallied, stayed out of the fray. The New Left was barely represented in the race—as the Harkin candidacy with its associations with organized labor and protectionism did not strike the appropriate political tone despite the populist rhetoric—and also had surprisingly limited appeal to the minority, feminist, and gay voters that sustained this segment of the Democratic party. After Harkin dropped out, the New Left mantle fell by default on Jerry Brown, but given his past record, it was impossible for him to arouse the fervor of the activists as George McGovern or Jesse Jackson had done.

Jackson's absence did Clinton a further favor by releasing the large black Democratic primary vote. This was an enormous advantage for Clinton, who, given his southern background, had by far the most experience in courting black voters of any candidate in the Democratic primary field, and it also ensured that Super Tuesday worked for the regional favorite as it had singularly failed to do in 1988. By adding the large black vote that had been unavailable to Al Gore in 1988 to the white southern conservative Democratic base, Clinton was able to trounce the northeastern neoliberal Tsongas throughout the region and amass a formidable lead in delegates (see Tables 6.3 and 6.4).

When the competition moved to the decisive midwestern battleground of Illinois and Michigan, Clinton was fortunate that his remaining opponents were Paul Tsongas and Jerry Brown. Tsongas had associated himself so strongly with fiscal conservatism and free trade that he alienated the large labor and black voting blocs in each state. Brown's countercultural associations mitigated the effects of his latter-day conversion to protectionism and the cause of organized Labor. Despite the occasional hiccup along the way, after Clinton's decisive victories in Illinois and Michigan, only a catastrophe could derail his campaign, and despite the high hopes of the Republicans and one or two prominent Democrats, it did not occur.

Clinton won, therefore, because of contingent factors: the vacuum on the New Left, the absence of a strong black candidate, the fact that he had the center-right field of the party all to himself, and the fact that in terms of fundamental ability and electoral appeal he stood head and shoulders above an extraordinarily weak field of opponents. Clinton's unique political background—the southern moderate who had opposed the Vietnam War, had worked for McGovern, and had an exemplary record on civil rights—also meant that unlike his major opponents he was acceptable to all factions of the party. In that sense he had the benefit of being everybody's second choice: The neoliberals would accept him over Jerry Brown if they could not have Tsongas, and the New Left infinitely preferred Clinton's vaguely populist rhetoric over Tsongas's fiscal stringency and associations with big business. For the same reasons, the remaining party regulars certainly preferred the Arkansas Democrat to the other two options as a presidential nominee.

The structural factors that had condemned to failure previous moderate and conservative candidacies had not disappeared. The activist base of the Democratic party, the powerful interest groups that participate in presidential nominating campaigns, and the basic Democratic presidential primary electorate remained to the left of center. And other things being equal, this situation still favors the more liberal candidate over the moderate one. But other things were not equal in 1992, because the liberal alternatives to Clinton were so weak and the absence of a serious black contender delivered the black vote to the Arkansas governor who, despite his more "conservative" position on the ideological scale, had more cultural affinity with black voters in his home region. In this respect Clinton's campaign was an echo of Jimmy Carter's in 1976.

One final factor that also made an impact in 1992 was the end of the cold war, which blurred the divisions in the Democratic party on national security that had formed the main basis for intraparty conflict since the Vietnam War and the 1968 election. Because those divisions were now far less relevant, it became easier for the party's neoliberal and New Left segments to accept a southern moderate. The fact that as a small-state governor Clinton had not been compelled to take strong stances on foreign and defense policy issues during the 1980s also worked to his advantage, whereas a southerner with a higher foreign policy profile, such as Sam Nunn, would have had more problems being accepted by the other sections of the party.

Whether Clinton could translate his success into a longer-term change in the dynamics of the presidential nominating politics is an issue that is discussed in the next chapter. In the next section we see how Clinton's task of uniting the Democratic party was made easier by the apparent disintegration of the Republican electoral coalition and, in addition, the emergence of a third major presidential candidate in 1992.

Bill Clinton's Democratic Party: The 1992 Democratic Convention and Perot Mark I

Clinton's prospects in the general election were enhanced considerably by the entry in the late spring of 1992 of Texas billionaire H. Ross Perot into the presidential race as an Independent candidate, because Perot's emphasis on the budget deficit and the "corrupt" state of American politics was directed at the Bush administration and also because his core constituency seemed to consist of upper-income, white male independents, a generally Republican-voting group since 1968.[49] Because Clinton trailed in third place behind Bush and Perot in some polls, this good fortune was not immediately apparent, but when the maverick Texan's poll ratings began to slide after a series of strategic misjudgements and public misstatements, followed by Perot's exit from the race during the Democratic National Convention in July, it became clear that the Democratic contender had been the prime beneficiary of the first Perot candidacy. Perot had served to energize voters who were dissatisfied with the political status quo, that is, George Bush. With Perot out of the race, for those voters who desperately wanted a change of leadership, the Democratic nominee, despite all his faults, was the only available vehicle.

Perot's withdrawal and Clinton's subsequent leap ahead of Bush in the polls constituted the culmination of an uncharacteristically successful Democratic National Convention. With the nomination effectively settled in early April, the Democrats for once had plenty of time to put together a platform behind which the various segments of the party could unite. Indeed, having learned the errors of previous campaigns, the Clinton campaign went out of its way to keep the platform as vague and unspecific as possible, rather than committing themselves to controversial policies for the sake of appeasing unreconciled segments of the party.[50] By keeping debate on the platform to a minimum, Clinton was able to turn the convention into a unity rally to launch his fall campaign.

Clinton also succeeded in keeping Rev. Jesse Jackson—whose appeasement had been a major problem for both the Mondale and the Dukakis campaigns—successfully in check. Indeed, Clinton took some pains to establish some distance between himself and Jackson, in the so-called Sister Souljah affair, concerning the intemperate remarks made by a rap singer with whom Jackson had shared a public platform. As far as the general election was concerned, establishing some distance from Jackson and the Democrats' New Left/minorities faction could do Clinton nothing but electoral good among the white, middle-class Reagan–Bush Democrats that he needed in order to secure victory in November.[51]

The most important decision left to Clinton before the convention opened was the selection of a vice-presidential nominee. As a southern moderate with neoliberal leanings, Clinton himself already covered much of the ideological ground within the party, but he pulled off another strategic coup by disregarding the conventional geographic ticket-balancing logic and asking his fellow "Rim-South," Baptist Democrat, Senator Albert Gore, Jr., of Tennessee, to be his running mate. Gore's selection reinforced the perception of the Democratic ticket as moderate, since the Tennessee senator was the most prominent Democrat who had voted in favor of the successful military operation against Saddam Hussein's Iraq in 1991 (although on some other issues such as the environment, Gore held positions quite far to the left of the nominee). Moreover, the selection of Gore put the Republicans even more on the defensive in the South—the base of their electoral college strength since 1968—while helping Clinton among Reagan–Bush Democrats in the crucial midwestern industrial states—Ohio, Illinois, Michigan, and Pennsylvania—that were culturally and geographically close to the Rim South base of the Democratic candidates and among environmentally concerned voters in the West. It was soon apparent that Gore was a net plus for the ticket in national terms, and his selection reinforced the domination of the DLC moderates over the national Democratic party in 1992.[52]

Clinton thus emerged from his national party convention considerably strengthened. Of course, Michael Dukakis had also enjoyed a lead in the national polls coming out of the 1988 convention, but he proved to be an utterly ineffectual candidate when exposed to the well-oiled Republican national machinery that had become so expert at utilizing so-called wedge issues to drive open the fissures within the Democratic party in presidential elections. Widespread dissatisfaction with the national status quo, the continuing economic recession, and a more politically astute candidate and national campaign team meant that the Republicans were already finding their task of discrediting the Democratic ticket somewhat harder than they had four years previously.

Perot Mark II and the Fall Campaign

Deciding to capitalize on the momentum established by the convention and Perot's withdrawal, the two Democratic nominees immediately began their fall campaign with a barnstorming "Meet the People" bus tour designed to establish a populist, "outsider" image likely to appeal to former Perot supporters. The enthusiastic popular response indicated that the venture was largely successful. As the economic news continued to be bad to indifferent, the Democratic campaign built up an ascendancy in the polls that it never lost for the remainder of the campaign.[53]

Clinton and his advisers had clearly learned the lessons of the disastrous Democratic campaigns of the 1980s. In contrast with Mondale's and Dukakis's emphasis on fiscal rectitude, Clinton largely ignored the deficit and concentrated on the hardships of the middle class, left behind in the Reagan boom and now squeezed by the economic recession.[54] Issues that were of particular concern to the white middle class, such as high health-care costs and education, were also emphasized. In times of recession, this message worked well and was constantly reiterated right up to polling day. The Democratic party is never stronger than when it can run on a populist economic ticket against an incumbent Republican administration during a recession (1932, 1960, 1976), and Clinton took maximum advantage of the opportunity by largely ignoring divisive social issues and foreign policy and concentrating on the economy. The celebrated sign at Clinton campaign headquarters, "It's the Economy, Stupid," perfectly captures the essence of a successful Democratic campaign.[55]

Bush and the GOP, of course, attempted the same strategy that had worked against Dukakis in 1988, and Clinton's more colorful past appeared to give them even more material with which to work. The Republican convention in Houston was deliberately planned to emphasize the old wedge issues of the 1980s, attacking the Democrats as the party of militant feminism, pacifism, gay rights, "quotas," and secularism, but in the context of the 1992 election, these attacks not only seemed irrelevant to the essential issues of the campaign, but they also made the GOP appear intolerant, exclusive, and shrill to the viewing public. The prominence given to Rev. Pat Robertson and erstwhile presidential candidate Patrick J. Buchanan only reinforced this perception.[56]

Ironically, the collapse of Soviet Communism and the subsequent end of the cold war, for which the Bush administration could take some credit, removed foreign policy — the Republican trump card in presidential elections since 1968 — as a significant issue in the campaign. Again and again, the debate came around to the issues of the economy and competitiveness, and Bush's efforts to stress his diplomatic achievements only served to reinforce the widespread popular perception that he was indifferent to the economic concerns of "middle America."[57]

Perhaps the final nail in the coffin of the Bush campaign was the reappearance of H. Ross Perot as a candidate in early October.[58] Initially the Republicans welcomed Perot's reentry, since they felt that it would upset Clinton's momentum and give them an opportunity to get back into the race.[59] In fact, the second Perot candidacy, though probably damaging Clinton to some extent, ultimately damaged Bush more for the same reasons as Perot's earlier campaign did in the spring: He provided an alternative for disillusioned Republicans who could not stomach Clinton and the Democrats; his anti-Washington rhetoric tended to be primarily directed against the incumbent administration; and the new voters that Perot attracted were those most unhappy with the political status quo and therefore unlikely to contemplate voting for the incumbent president. Moreover, Perot's appearances in the three televised presidential debates undoubtedly hurt Bush, since they moved the focus of those encounters away from the direct Bush–Clinton confrontation that the incumbent needed in order to try to upset the Clinton bandwagon at this late stage in the campaign. Perot's folksy populism detracted from Bush's efforts to concentrate his fire on Clinton, and the president found himself under attack from both sides.[60]

Clinton meanwhile adhered to his carefully honed populist/competitiveness message, and this, together with his evident composure and polished delivery, enabled him to emerge from the debates in a stronger position than ever. Despite some late blips in the polls, Clinton's electoral college position appeared to be impregnable, with a solid anchor of support in New York, the Pacific states including California, and the states carried by Dukakis in 1988, while also being competitive with Bush almost everywhere else, including the post-1968 Republican electoral base of Florida, Texas, and the Deep South.[61] Now, in contrast with the situation during the 1980s, it was the GOP that had to spread its resources thinly in a desperate attempt to assemble a plausible electoral college majority in the last days of the campaign. The wisdom of an all-southern Democratic ticket was never more clearly demonstrated than at the tail end of the 1992 election.

The Election Result and the New Democratic Administration

The result of the general election was hardly surprising given the course of the campaign. Clinton won by a considerable margin of 370 to 168 in the electoral college, and by the much closer margin of 43 percent to 38 percent for Bush and 19 percent for H. Ross Perot in the popular vote. At 55 percent, the national turnout for this election was the highest since 1968, largely because of the impact of the Perot campaign.[62]

In electoral college terms, Clinton's win was impressive. He carried thirty-two states plus the District of Columbia, including every state that mattered outside the South. In his native region, Clinton's showing was the best of any Democrat since Carter. He won Arkansas, Tennessee, Louisiana, and Georgia and came very close in Florida and North Carolina. Compared with Dukakis he raised the Democratic vote marginally but significantly in most of the southern states (see Table 6.6).

Clinton also did marginally better than Dukakis had among white southerners and born-again Christians, although both categories still went Republican by large margins. Among Catholics, union members, and male voters, all of whom had deserted the Democrats in large numbers during the 1980s, Clinton did marginally less well than Dukakis had in 1988, but the prolonged recession and the apparent moderation of the Clinton–Gore ticket on social issues kept enough of them in line for the Democrat to prevail in a three-way contest. With a national popular vote three points lower than that of his Democratic predecessor in 1988 and only marginal advances at best among the key swing-voting groups, Clinton's triumph thus was tenuous and rather limited.[63]

Table 6.6. Democratic Vote in the Southern States, 1988 and 1992 (percent)

State	Dukakis 1988	Clinton 1992	Difference 1988–92
Alabama	40	42	+ 1
Arkansas	42	54	+ 12
Florida	39	39	0
Georgia	40	44	+ 4
Louisiana	44	46	+ 2
Mississippi	39	41	+ 2
North Carolina	42	43	+ 1
South Carolina	38	40	+ 2
Tennessee	42	47	+ 5
Texas	43	37	− 6
Virginia	39	41	+ 2

Sources: Rhodes Cook, *Race for the Presidency: Winning the 1992 Nomination* (Washington, DC: Congressional Quarterly Press, 1991), p. 113; and *CQ Guide to Current American Government* (Washington, DC: Congressional Quarterly Press, 1993), p. 15.

With major damage from Ross Perot, the Republican vote for president sank precipitously both in the South and nationwide, although Perot was noticeably weaker in the South (largely due to his poor showing among black voters) than elsewhere in the nation. It appeared that in the South, social conservatism still cut deeply enough to sustain a narrow Republican plurality, though greatly weakened in comparison with that four and eight years previously. Of course, it is all too easy to attribute Clinton's victory to Perot and the Republican voters that he wooed away from Bush, yet the postelection evidence concerning the Perot voters' second choices indicates that they would have marginally voted for Clinton in a straight contest, so the Democratic victory cannot be merely written off as a fluke outcome created by the Perot candidacy.[64]

Assuming that Perot is not to become a permanent fixture on the American political landscape, this still leaves Clinton with the task of winning over enough of the Perot vote to establish a new presidential majority for the Democratic party. To do that, in addition to bringing some tangible economic benefits to the "forgotten middle class" that won him the election, Clinton must maintain the delicate ideological balancing act that was so successful in the 1992 campaign. The politically moderate-to-conservative white middle class and the "minorities"—blacks, Latinos, feminists, gays—must be kept in alliance for the Clinton administration to be successful, yet their agendas still conflict significantly, particularly with regard to social issues like affirmative action and gay rights.

An indication of Clinton's problems in this direction is his commitment to end the policy of indiscriminately discharging from the armed forces persons who are revealed to be gay. This commitment aroused little attention during the campaign, but it undoubtedly (with more than a little assistance from the Republican National Convention) helped Clinton and the Democrats mobilize a significant number of gay voters on their side, in major states like New York and California where much of the gay population is concentrated.[65] After the election, General Colin Powell, chairman of the Joint Chiefs of Staff, and Senator Sam Nunn, chairman of the Senate Armed Services Committee, made it clear that they were opposed to such a measure. An immediate outcry from socially conservative voters on this issue, combined with mixed signals from the Clinton White House, permitted the "gays in the military" issue to dominate the first week of the Clinton presidency. Ultimately the new president was compelled by political exigencies to retreat somewhat from his commitment, stating that the ban would be rescinded only after a six-month delay and following consultation with the military and Senate hearings on the issue. The lesson from the "gays in the military" fracas appeared to be that unless such sensitive social issues remain suppressed by economic concerns in the media and in the minds of the electorate, Clinton's chances of holding onto the socially conservative Democrats who had supported him in 1992 and winning over the crucial bloc of Perot voters to his side in 1996 are rather limited.[66]

Clinton's cabinet choices again reflected his need to placate the swing voters in the middle class without alienating the powerful interest groups in the Democratic party that had been waiting for twelve long years for a share of federal government power. Having emphasized the economy during the campaign, Clinton made one of the pillars of the Texas business establishment and the southern conservative wing of the Democratic party, Senator Lloyd Bentsen, his secretary of the treasury. For the "politically correct," Clinton's cabinet choices included six women (including America's first female nominee for attorney general, Zoe Baird), four blacks, and two Latinos, most of them in the lower-ranking domestic policy departments.[67] With the exception of Baird, the nominees to the four major departments — Warren Christopher (State), Les Aspin (Defense), and Bentsen (Treasury) — were white males of generally moderate political bent.

The selection process did not take place without one or two hiccups. About midway through the appointments, women's groups, led by the National Organization of Women (NOW), made it quite clear that they felt Clinton was not appointing enough women to senior cabinet positions. As a consequence it appears that Clinton changed his nominee for secretary of energy from retiring Senator Tim Wirth of Colorado (a white male) to utility company executive (and black female) Hazel O'Leary. After a calculated outburst at a press conference about "bean counters" and "quotas," Clinton also took some pains to find another woman as his nominee for the post of attorney general rather than the candidate specifically promoted by the women's groups.[68] In the foreign policy field, there was some discontent among neoconservative Democrats who had been wooed back into the Democratic ranks during the 1992 campaign in the belief that Clinton would intervene more energetically to promote democracy around the world than the Republicans had, in that the president-elect saw fit to fill most of the top foreign policy positions with veterans of the Carter administration: Christopher (secretary of state), Anthony Lake (national security adviser), and Madeleine Albright (ambassador to the United Nations).[69]

In general, the Clinton cabinet seemed to be largely bereft of personnel (with the exceptions of Lloyd Bentsen and the education secretary, former South Carolina Governor Richard Riley) from the DLC/southern conservative wing of the party, and it was ostensibly leaning more toward the "bean counters": the traditional Democratic interest groups and the left wing of the party. However, the presence of Clinton and Gore at the head of the administration was far more visible to the public at large and the key swing-voting groups than were the cabinet officers, and perhaps cabinet appointments were a small price to pay to the Democratic party establishment in return for acquiescence to Clinton's bolder measures to aid the middle class and restore national competitiveness.

Conclusion

Bill Clinton disrupted the original plan of this book by having the temerity to win a presidential election just as I was developing a hypothesis that (1) because of structural factors a self-proclaimed "moderate" was unlikely to win a major party nomination and furthermore (2) was particularly unlikely in the Democratic party. At least, Clinton's November victory verified the corollary of my original hypothesis, that if the

Democrats ever did nominate a moderate, and particularly a southern
moderate, their electoral chances would be greatly enhanced. The failure
of my theory to work in 1992 serves as a salutary reminder of the
limitations of all "theories" in political science when they come up
against the Machiavellian qualities of *fortuna* and *virtu*, both of which
operated in favor of Bill Clinton in 1992. Clinton saw his good fortune in
the shape of a very weak field of Democratic presidential candidates and
in an incumbent with greater weaknesses than were apparent to many
other observers early in the campaign, and he brilliantly converted those
advantages into eventual success.

Whether the change in the balance of forces within the Democratic
party, with the DLC and southern moderates in the ascendant, is permanent is another issue, largely contingent on how effectively the Clinton
administration performs in office. What the renaissance of the center in
the Democratic party implies for American politics generally is discussed in the concluding chapter.

7

Conclusion: The Future of the Southern Democrats and American Party Factionalism in the 1990s

Given the spectacular success of the all-southern Democratic Clinton–Gore ticket in 1992, the short-term prospects for the southern wing of the Democratic party would appear to be very bright indeed. In defiance of the apparent logic of the presidential nominating process during the 1968–88 period, an ostensibly moderate southerner was able to win the Democratic presidential nomination with ease and end more than a decade of Republican domination of the presidency. Whereas from the Civil War up to almost the end of the New Deal era it was virtually inconceivable that the Democratic party would nominate a southerner for president, since 1964 the only times when the Democratic party has won presidential elections has been with southern nominees: Johnson, Carter, and now Clinton. Yet parties exist not only to win elections but also to govern in accordance with a set of principles or policies, and the reemergence of the southern wing of the Democratic party needs to be confirmed by a successful Clinton presidency followed by reelection in 1996. Failure in office for a second time might well doom the southern conservative wing of the party to a further spell in the wilderness while the Republicans dominate the White House.

Southern Democrats and the Clinton Presidency

Clinton faces the same problem that confronted Carter: how to consolidate the party's support among the largely white, middle- to working-class, suburban voters who will respond to populist economics in bad times but who are generally nationalistic and socially conservative, without alienating the party's post-1968 activist base (the people who provide the money, the time, and much of the labor for Democratic campaigns) which is still strongly committed to the agenda of late-1960s social and cultural liberalism. This is not to imply that the American middle class are generally bigoted rightists, any more than that Democratic activists are necessarily sex-obsessed guardians of political correctness. Nevertheless, as Clinton discovered as early as the presidential transition, it is hard to appear to be satisfying the claims of one group without arousing suspicion from the other. What Clinton must do is ensure that the cultural cleavage that has dominated American politics since 1968 can be suppressed by a new electoral cleavage around New Deal–like "haves-versus-have-nots" issues. One great advantage that he possesses is the end of the cold war, which had served as the major area of factional strife within the Democratic party since the Vietnam War and which ultimately (in its different manifestations) proved to be the electoral undoing of both Johnson and Carter. Yet the Democrats remain as uncertain as the Republicans are about what kind of American foreign policy is appropriate to a post-Communist world, and it is not inconceivable that new divisions might open up among the party factions on isolationist-versus-interventionist lines during the 1990s.

Despite Clinton's triumph, the southern wing of the party still faces several major disadvantages in the intraparty factional competition at the national level. For a start, the gradual southern realignment in favor of the GOP continued in 1992. Even with two white southerners at the top of the national Democratic ticket, Clinton gained only 34 percent of the southern white vote (according to the VRS survey) and 27 percent of the electoral votes of his home region and lost the two southern megastates of Texas and Florida. The Republicans gained more U.S. House seats owing to the effects of the 1991 congressional reapportionment and redistricting, which produced more minority districts and also (because of the loss of the core Democratic black vote) more Republican districts (see Table 7.1). At the state legislative level (where redistricting also took place) a similar pattern emerged. Although the South has won back the national Democratic party, the Democrats are gradually becoming weaker in the South. If this trend persists, it will continue to reduce the South's influence in Congress and the national party.

Table 7.1. Redistricting and the 1992 U.S. House Elections in the South

State	Before Redistricting		1992 Outcome		GOP
	R	D	R	D	Gain/Loss
Alabama	2	5	3	4	+1
Arkansas	1	3	2	2	+1
Florida	10	9	13	10	+3
Georgia	1	9	4	7	+3
Louisiana	4	4	3	4	−1
Mississippi	0	5	0	5	0
North Carolina	4	7	4	8	0
South Carolina	2	4	3	3	+1
Tennessee	3	6	3	6	0
Texas	8	19	9	21	+1
Virginia	4	6	4	7	0
South	39	77	48	77	+9

Source: *Southern Political Report*, 10 November 1992.

In regard to Capitol Hill, nothing that occurred at the polls in 1992 served to enhance the influence of the South. The crucial leadership positions remained in the hands of northern liberals, and the Republicans have two southerners in prominent leadership positions: Congressman Newt Gingrich (Georgia) as minority whip (the second-ranking position) in the House and Senator Thad Cochran (Mississippi) as conference chairman (the third-ranking position) in the Senate. In the only significant Democratic leadership contest after the 1992 elections, the South lost again when Congressman John Spratt of South Carolina was defeated, 149 to 112, by the more liberal Congressman Martin Olav Sabo of Minnesota, for the chairmanship of the Budget Committee. Like Jimmy Carter, Clinton will have to work with a Congress dominated by his own party but certainly not by his own faction. The promised reforms in areas like health care and education, or reductions in spending as part of a deficit-reduction package, will be difficult for the new administration to pass in the face of powerful liberal interest groups with congressional allies. On the other hand, a Clinton presidency calculated to appeal to the mainstream of the congressional Democratic party may run the risk of alienating the southern Democrats and encourage the reemergence of the conservative coalition as a major force in Congress.

In short, the long-term problems facing the southern faction of the Democratic party were not resolved by the Clinton victory in 1992. The effects of the New Deal realignment, the civil rights movement, the downfall of the committee seniority system in Congress, and the movement of the presidential nominating process away from the party leaders and the convention to the new presidential elite of issue activists, national interest groups, and the news media all worked against the interests of the party's southern wing. By changing the context of the party alignment from one based on region, ethnicity, and culture to one based on class, the New Deal isolated the southern Democrats decisively from their national brethren and made them an aberrant reactionary element (and ultimately a dispensable element) in an otherwise "center-left" electoral coalition. Then the civil rights revolution isolated the traditional Democratic South even more and assisted the creation of a viable southern Republican party in which the most conservative elements of the old Democratic South found a comfortable home. The obstruction of civil rights, together with the expansion of Democratic party support outside the South in the wake of the New Deal, discredited the South's domination of Congress through the committee seniority system and the conservative coalition and ultimately led to the more atomistic but also more partisan congresses of the 1970s and 1980s, in which southern influence, although intermittently present on certain issues, was greatly reduced. Finally, the movement of the selection of the Democratic presidential nominee from the national convention to the primaries undermined the regional cohesion through which traditional southern Democratic elites had exercised influence in the national party. Southern Democratic party leaders were thus forced by the dynamics of the reformed nominating process to endure the nomination of candidates such as McGovern, Mondale, and Dukakis, who were least likely to appeal to the vast majority of southern voters.

Thanks to the very effective mobilization of the southern black vote after the 1965 Voting Rights Act and the greater degree of racial moderation in the South (due largely to the long-term social effects of migration, economic development, and the "nationalization" of politics in general) after the civic turmoil of the 1950s and 1960s, the Democratic party in the South was able to reconstruct itself as the "have-nots" party in a southern two-party system loosely based on class politics, just as V. O Key and his contemporaries had hoped during the 1940s. Deep-seated historical and cultural ties and the association of the southern GOP with a still relatively narrow social stratum of migrants and big businessmen allowed the Democrats to remain dominant in congressional, state, and local elections, although by the late 1960s in senatorial and gubernatorial races, the southern GOP had become an equal competitor in most southern states. What Key did not anticipate was the development of a split-level electoral alignment in the United States after 1968, in which the Republicans dominated the presidency and the Democrats every other level of American politics.

The key to the split-level electoral realignment was the South. In subpresidential contests that focused on the protection of governmental services and constituent representation, southern voters still tended— other things being equal—to gravitate toward Democratic candidates. In presidential contests, issues of national security and cultural symbolism came to be decisive in the wake of the Vietnam War and the 1960s revolution in public morals. These issues polarized the intraparty factions of the Democratic party so severely that except in the very special circumstances of 1976, broad-based agreement on a presidential candidate and party platform was all but impossible. Moreover, the nominating process, skewed toward the New Left and neoliberal sections of the party by the influence of single-issue activists, strident national interest groups, and the impact of unrepresentative northern states early in the process, meant that the Democrats tended to end up with candidates who were the least likely to have major electoral appeal in southern states outside the black community.

Writing off the South at the outset also made it extremely hard for those Democratic candidates to put together anything like a majority of electoral college votes. Carter won the 1976 nomination because of his "outsider" appeal and despite, rather than because of, his associations with the South, yet it was the solidity of his southern support that enabled him to pull off a victory in the general election. Once in office, his "southernness" was not an asset in dealing with the largely nonsouthern Democratic Congress, and the popular identification of his administration with cultural liberalism and ineptitude in foreign affairs lost him the South and the presidency in 1980. In 1984, the conservative Democratic contenders—Ernest Hollings, Reuben Askew, and John Glenn—were left behind in the bitter struggle among the old northern New Deal regulars (Walter Mondale), the post-1968 neoliberals (Gary Hart), and the New Left and the minorities (Jesse Jackson), and the South appeared to be almost irrelevant to the intraparty struggle. In the wake of the 1984 debacle, southern party leaders attempted to organize their forces for 1988 by creating the regional Super Tuesday primary, but the ploy backfired because of the primary calendar and the impact of the New

Hampshire primary, Jesse Jackson's solid support from southern black voters, and the southerners' own failure to organize behind a single credible candidate. Again the South found itself marginalized in a New Left/neoliberal conflict between Dukakis and Jackson, neither of whom was palatable to the southern general electorate as a potential president.

Bill Clinton in 1992 defied the longer-term structural factors that operate against any southern Democratic presidential aspirant, by exploiting the context of the election and the ineptitude of his opponents to grab the presidential nomination. In the fall he was able to win an overwhelming victory in the electoral college while only winning four of the eleven southern states, but his greater regional appeal than either Mondale or Dukakis offered succeeded in tying up the Republican campaign in what, since 1968, had been their surest electoral college states and enabled Clinton to maximize the discontent with Bush in the rest of the country.

Having lost clout on Capitol Hill with the end of the committee seniority system, southern Democrats in the House and Senate found themselves drawn closer ideologically to their northern Democratic colleagues, but becoming an increasingly marginal section of the party. The slow but steady advance of the southern GOP reduced their influence in the House caucus, and it became more and more difficult during the 1980s for a southern Democrat to win and maintain a position in the party leadership. The southern conservative members were still the decisive element in both Houses on many of the most contentious issues ("Reaganomics," Bork, Tower, Thomas, the Gulf War), but they had increasingly less control over the direction of their party. Moreover, the prospects of further drastic redistricting in the southern states in order to produce more minority U.S. House districts portended the continuing gradual diminution of white southern influence in the Democratic party on Capitol Hill.

None of these longer-term factors has been reversed or even arrested by Clinton's victory in 1992. Clinton is also far from being a stereotypical southern Democrat. His (and Gore's) political position is well to the left end of the southern Democratic ideological spectrum, and he could not have been nominated without establishing close ties to neoliberal Democrats and party regulars outside his home region. Looking to Clinton to overhaul the Democratic party organizationally and change its ideological orientation may also be misguided. Contemporary American political parties are not highly structured and disciplined organizations but, rather, are elusive and amorphous entities, incapable of being turned around by even a powerful incumbent president. Despite the pre-1992 speculations of the DLC's Al From and some congressional moderates, it is unlikely that Clinton will be any more interested than Jimmy Carter was in trying to make fundamental changes in a nominating process that worked so well for him in 1992.

A successful Clinton presidency would undoubtedly arrest for a time the slow long-term decline in the South's influence over the national Democratic party. But barring a truly major sea change in American electoral alignments during the 1990s, that decline is unlikely to be reversed.

A New Factional Pattern?

The Clinton nomination by the Democrats in 1992 was clearly at odds with the general pattern of American party factionalism set out in the first chapter, so it is appropriate in this final section to discuss whether that pattern is still a useful analytical tool for studying American intraparty conflict or whether it needs to be substantially revised to take account of new developments in the 1990s, such as the end of the cold war and the onset of widespread national concern about the efficacy of the American political process.

The earlier account of contemporary American party factionalism stated that in contrast with the previously accepted scholarly wisdom on the subject, fairly clear and established party factions based largely on differences over political ideology and style have been present in American national parties since the 1930s. Social and technological changes (the decline of the party machines, the growing white-collar sector, the advent of national news media) encouraged the development of polarized factions in both parties, and patterns of factional conflict have recurred consistently over the past quarter-century in contests for the presidential nomination and for seats in Congress. Nothing that occurred in 1992 contradicted these assertions.

But I also argued that structural factors in modern American politics (primary turnout, sources of funds and labor power, media criteria for analysis) tended to favor those factions in each party that were more concerned with ideology and the need to emphasize it, as opposed to the more centrist factions: the liberal Republicans and the conservative (mainly southern) Democrats. Again there was confirmation of this pattern in 1992. The incumbent Republican president was faced with a serious primary challenge from the far right of the party, in the shape of Patrick J. Buchanan, who, although he did not come near to winning a single primary election, forced the Republican party to stage a stridently right-wing national convention in order to accommodate him.

What was different in 1992 was that as in 1976, the only other election since 1968 that does not fit the pattern, the Republicans were the party that suffered more from prolonged factional conflict than the Democrats did. Between 1968 and 1988, Republican factional conflict was generally not as bitter as it was in the Democratic party, because the liberal wing of the GOP had been more or less eliminated as a significant force by the early 1970s, and there was a remarkable degree of ideological homogeneity among Republicans on issues such as a tough anti-Communist foreign policy, free markets, and limited government. There were significant divisions on cultural issues between more libertarian and authoritarian conservatives, but these were largely submerged, as Republicans found it easy to unite against the very liberal Democratic nominees of the 1970s and 1980s. Moreover, Republican rules for allocating delegates tended to favor early primary winners and thus produce a nominee very early in the process who had plenty of time to reconcile the party before the convention.

In 1992 Clinton and the Democrats got lucky in this sense, because the poor quality of the opposition and Clinton's unique ability to straddle the party's factional lines helped make him an early consensus nominee. But there was more to his victory than fortune, and it may be that the most significant change in the pattern of American intraparty conflict in the 1990s is that conflict will become as strident (if not more so) within the Republican party as it had been among the Democrats during the 1970s and 1980s. The collapse of Soviet Communism deprived the Republicans of the issue that most effectively united them and disunited the Democrats, and in the absence of Communism, economic and cultural issues have come to the fore. Despite the 1980s boom, economic issues continued to work well for the Democrats, enabling them to maintain their control of Congress while losing presidential elections by landslide margins. In 1992, with a presidential election campaign being conducted during a prolonged economic recession, the Democrats finally had an opportunity to win the presidency.

Cultural issues were exploited by the Republicans during the 1980s against Democratic nominees who stood well to the left of popular opinion on those issues, but there was always a great deal of unease in large sections of the Republican electoral coalition over the GOP's adoption of the moral agenda of the religious right, particularly on the abortion issue. With the defeat of Communism, the abortion conflict broke out into the open, and it partly contributed to the rise of the Perot candidacy, which attracted many discontented, economically conservative but socially moderate-to-liberal Republicans. The anxieties of these libertarian Republicans were confirmed by the GOP convention and the initial "family values" theme of the fall campaign.

Economic recovery and the avoidance of major foreign policy disasters under Clinton will probably exacerbate this conflict within the Republican party and give the Democrats a chance to grab the center and redefine the nature of the American party system, as the Republicans did in presidential politics after 1968. Should this happen, it will be within Republican ranks that the most intense factional conflict will take place, and because of these factors, this conflict may well be won by the "religious right." If this is the case, Democrats will find it easier to unite against Republican candidates of the far right wing, who are seen as being out of the national mainstream of early twenty-first-century America, just as the Republicans found it easier to unite against very liberal Democrats who were out of the mainstream of 1970s and 1980s America.

A failed Clinton presidency, on the other hand, would probably reestablish the post-1968 pattern. Like Carter, Clinton would suffer a strong intraparty challenge, probably from the New Left faction of the party, and the Republicans, sensing weakness, would find it easy to unite under a candidate acceptable to both the authoritarian and libertarian factions of the party, such as Jack Kemp or Richard Cheney, who would be able to grab back the middle ground from another failed Democratic presidency.

In any event, the intraparty factions and patterns of factional conflict over ideology described in the opening chapter are likely to persist into the next century. What might change after 1992 is the balance between the two parties in presidential elections if the Clinton presidency is able to create a new national consensus on Democratic issues, particularly on the economy. Yet though that might restore the Democratic dominance of the New Deal era, the party factions of the post-1968 period will endure, and unless the structure of the nominating process is fundamentally moved away from primary elections, the more "ideological" or "extreme" faction in each party will probably prevail.

NOTES

Chapter 1

1. Gordon S. Wood, *The Creation of the American Republic, 1776–1787* (New York: Norton, 1972), pp. 46–90.

2. Isaac Kramnick, ed., *Viscount Bolingbroke: Political Writings* (New York: Appleton-Century-Crofts, 1970).

3. Ibid., p. 76.

4. *The Federalist Papers*, nos. 10 and 51.

5. Giovanni Sartori, *Parties and Party Systems: A Framework for Analysis*, vol. 1 (Cambridge: Cambridge University Press, 1976), pp. 4–38.

6. Ibid., pp. 71–115.

7. Richard Rose, *The Problem of Party Government* (London: Macmillan, 1974), pp. 320–21.

8. Sartori, *Parties and Party Systems*, pp. 71–115.

9. For a good comparative discussion of party factionalism, see Raphael Zariski, "Party Factions and Comparative Politics: Some Empirical Findings," in Dennis C. Beller and Frank P. Belloni, eds., *Faction Politics: Political Parties and Factionalism in Comparative Perspective* (Santa Barbara, CA: ABC–Clio, 1978), pp. 19–38.

10. See ibid.; and Dennis C. Beller and Frank P. Belloni, "Party and Faction: Modes of Political Competition," in Beller and Belloni, eds., *Faction Politics*, pp. 417–48.

11. Thomas H. Roback and Judson L. James, "Party Factions in the United States," in Belloni and Beller, eds., *Faction Politics*, pp. 329–55. Note that Judson and James also introduce yet another term, *wing*, into the discussion.

12. Howard L. Reiter, "Intra-Party Cleavages in the United States Today," *Western Political Quarterly* 34 (1981): 287–300.

13. See Nicol C. Rae, *The Decline and Fall of the Liberal Republicans: From 1952 to the Present* (New York: Oxford University Press, 1989), pp. 10–45; and Earl Black and Merle Black, *The Vital South: How Presidents Are Elected* (Cambridge, MA: Harvard University Press, 1992), pp. 79–115.

14. Barbara Sinclair, *Congressional Realignment: 1925—1978* (Austin: University of Texas Press, 1982), pp. 18–50; and Richard Franklin Bensel, *Sectionalism and American Political Development: 1880—1980* (Madison: University of Wisconsin Press, 1984), pp. 60–174.

15. See Paul Kleppner, "Partisanship and Ethnoreligious Conflict: The Third Electoral System, 1853–1892," in Paul Kleppner et al., eds., *The Evolution of American Electoral Systems* (Westport, CT: Greenwood Press, 1982), pp. 113–47; and David Hackett Fisher, *Albion's Seed: Four British Folkways in America* (New York: Oxford University Press, 1989), pp. 783–898.

16. See Bensel, *Sectionalism and American Political Development*, Preface and pp. 3–59.

17. On the East/West tensions in the GOP and in Republican progressivism, see Rae, *The Decline and Fall of the Liberal Republicans*, pp. 10–25.

18. Ibid., pp. 10–45.

19. On the influence of the South in the pre–New Deal Democratic party, see V. O. Key, Jr., *Southern Politics in State and Nation* (New York: Knopf, 1950), pp. 315–82; Dewey W. Grantham, *The Life and Death of the Solid South: A Political History* (Lexington: University Press of Kentucky, 1988), pp. 58–77; Sinclair, *Congressional Realignment*, pp. 18–50; and Black and Black, *The Vital South*, pp. 79–115.

20. I should add, however, that single-party dominance in many state and local governments and the development of political machines during the late nineteenth century did give rise to classic instances of clientelistic party factionalism in American state and local government.

21. On party factions in the New Deal era, see Bensel, *Sectionalism and American Political Development*, pp. 104–174; James T. Patterson, *Congressional Conservatism and the New Deal: The Growth of the Conservative Coalition in Congress* (Lexington: University of Kentucky Press, 1967), pp. 32–76; and Frank Munger and James Blackhurst, "Factionalism in the National Conventions, 1940–1964: An Analysis of Ideological Consistency in State Delegation Voting," *Journal of Politics* 27 (1965): 375–94.

22. See Key, *Southern Politics*, pp. 345–82; Donald R. Matthews, *U.S. Senators and Their World* (New York: Vintage Books, 1960), pp. 92–117, 162–66; and William S. White, *Citadel: The Story of the U.S. Senate* (New York: Harper Bros., 1956) pp. 67–79, 179–97, 199–211.

23. On Republican factional conflict, see Rae, *The Decline and Fall of the Liberal Republicans*, pp. 25–45.

24. James Q. Wilson described the conflict between the reformers and the old New Dealers as one between two distinct types of party activists: "amateurs" and "professionals." See James Q. Wilson, *The Amateur Democrat: Club Politics in Three Cities* (Chicago: University of Chicago Press, 1962), pp. 1–31. On the Republicans, see Aaron Wildavsky, *The Revolt Against the Masses: And*

Other Essays on Politics and Public Policy (New York: Basic Books, 1971), pp. 246–69.

25. See Wildavsky, *The Revolt Against the Masses*, pp. 246–69.

26. See Rae, *The Decline and Fall of the Liberal Republicans*, pp. 196–214.

27. See Jon F. Hale, "The Institutionalization of a Party Faction: The Case of the Democratic Leadership Council," paper presented at the annual meeting of the Midwest Political Science Association, Chicago, 18–20 April, 1991.

28. Rae, *The Decline and Fall of the Liberal Republicans*, pp. 1–45.

29. Kevin Phillips's *The Emerging Republican Majority* (Garden City, NY: Anchor Books, 1970), which argued for the eventual merging of the Nixon and Wallace constituencies after the 1968 election, served as the electoral handbook for the Republican Right during the 1970s.

30. Rae, *The Decline and Fall of the Liberal Republicans*, pp. 1–45.

31. Ibid., pp. 46–77.

32. Ibid., pp. 157–95.

33. The tension among the regulars, the New Left, and the neoliberals in the Democratic party also reflects a broader tension between modernizing and traditional cultures in American society, which was anticipated by James Q. Wilson in the early 1960s. See Wilson, *The Amateur Democrat*, pp. 258–370. See also James Davison Hunter, *Culture Wars: The Struggle to Define America* (New York: Basic Books, 1991), pp. 1–51; Everett Carll Ladd, Jr., with Charles D. Hadley, *Transformations of the American Party System*, 2nd ed. (New York: Norton, 1978), pp. 181–274; Jonathan Rieder, *Canarsie: The Jews and Italians of Brooklyn Against Liberalism* (Cambridge, MA: Harvard University Press, 1985), pp. 1–9, 233–63; Christopher Lasch, *The True and Only Heaven: Progress and Its Critics* (New York: Norton, 1991), pp. 476–532.

34. On the Southern *volte-face* on civil rights in Congress, see Charles S. Bullock, III, "The South in Congress: Power and Policy," in James F. Lea, ed., *Contemporary Southern Politics* (Baton Rouge: Louisiana State University Press, 1988), pp. 177–93.

35. On the upper- middle–class, white-collar bias of Democratic primary electorates, see Nelson W. Polsby, *Consequences of Party Reform* (New York: Oxford University Press, 1983), pp. 157–67.

36. On the "momentum effect" of the New Hampshire primary, see William G. Mayer, "The New Hampshire Primary: A Historical Overview," in Gary R. Orren and Nelson W. Polsby, eds., *Media and Momentum: The New Hampshire Primary and Nomination Politics* (Chatham, NJ: Chatham House, 1987), pp. 9–37.

37. Bullock, "The South in Congress," pp. 177–93.

38. See Hale, "The Institutionalization of a Party Faction." Although established by southern conservatives, in recent years the DLC has acquired an increasingly neoliberal cast.

39. On the CDV, see Robin Toner, "Liberals in Search of Values Run into Discord over War," *New York Times*, 27 January 1991, p. 11.

40. On the Ripon society, see Rae, *The Decline and Fall of the Liberal Republicans*, pp. 79–88, 205–14. On the conservative network, see Sidney Blumenthal, *The Rise of the Counter-Establishment: From Conservative Ideology to Political Power* (New York: Harper & Row, 1988).

41. Byron E. Shafer, "The Notion of an Electoral Order: The Structure of Electoral Politics at the Accession of George Bush," in Byron E. Shafer, ed., *The End of Realignment? Interpreting American Electoral Eras* (Madison: University of Wisconsin Press, 1991), p. 61

Chapter 2

1. John Crowe Ransom, "Reconstructed but Unregenerate," in "Twelve Southerners," *I'll Take My Stand: The South and the Agrarian Tradition* (Baton Rouge: Louisiana State University Press, 1977), pp. 26–27.

2. David Hackett Fischer, *Albion's Seed: Four British Folkways in America* (New York: Oxford University Press, 1989), pp. 207–418.

3. Ibid., pp. 13–205. See also Daniel J. Boorstin, *The Americans: The Colonial Experience* (New York: Vintage Books, 1958), pp. 3–31, 97–143.

4. Fischer, *Albion's Seed*, pp. 605–782. See also James G. Leyburn, *The Scotch-Irish: A Social History* (Chapel Hill: University of North Carolina Press, 1962).

5. See Fischer, *Albion's Seed*, pp. 405–18, 772–82. W. J. Cash, *The Mind of the South* (New York: Vintage Books, 1941), pp. 1–60, emphasizes the Scotch-Irish "backcountry" culture as being dominant in the South and the "southern aristocrat" as being largely mythical. For a contrasting account that emphasizes the "gentlemanly" tradition, see Richard M. Weaver, *The Southern Tradition at Bay: A History of Postbellum Thought* (Washington, DC: Regnery-Gateway, 1989), pp. 31–95.

6. Cash, *The Mind of the South*, p. 52.

7. Ibid., p. 587.

8. Bertram Wyatt-Brown, *Southern Honor: Ethics and Behavior in the Old South* (Oxford: Oxford University Press, 1982), pp. 1–2. On the concept of honor, see also Gordon S. Wood, *The Radicalism of the American Revolution* (New York: Knopf, 1992), pp. 24–42.

9. See Joel H. Silbey, "Parties and Politics in Mid-Nineteenth Century America," in Joel H. Silbey, *The Partisan Imperative: The Dynamics of American Politics Before the Civil War* (New York: Oxford University Press, 1985), pp. 33–49; and J. Mills Thornton, III, "Jacksonian Democracy," in Charles Reagan Wilson and William Ferris, eds., *Encyclopedia of Southern Culture* (Chapel Hill: University of North Carolina Press, 1989), pp. 629–31.

10. On the unraveling of the Jacksonian party system, see James L. Sundquist, *Dynamics of the Party System: Alignment and Realignment of Political Parties in the United States*, rev. ed. (Washington, DC: Brookings Institution, 1983), pp. 50–105.

11. On Reconstruction's impact on the white southerners' political consciousness, see Cash, *The Mind of the South*, pp. 105–47.

12. Ibid., pp. 148–71. On the post–Civil War southern economic system, see Richard K. Scher, *Politics in the New South: Republicanism, Race and Leadership in the Twentieth Century* (New York: Paragon House, 1992), pp. 23–59; and Edward L. Ayers, *The Promise of the New South: Life After Reconstruction* (New York: Oxford University Press, 1992), pp. 104–31, 187–213.

13. On segregation, see C. Vann Woodward, *The Strange Career of Jim Crow*, 3rd rev. ed. (New York: Oxford University Press, 1974), pp. 31–111; see also Ayers, *The Promise of the New South*, pp. 132–59.

14. See J. Morgan Kousser, *The Shaping of Southern Politics: Suffrage Restriction and the Establishment of the One-Party South, 1880–1910* (New Haven, CT: Yale University Press, 1974), pp. 11–44; Dewey W. Grantham, *The Life and Death of the Solid South: A Political History* (Lexington: University Press of Kentucky, 1988), pp. 8–11; and Ayers, *The Promise of the New South*, pp. 34–54.

15. On southern populism, see Grantham, *The Life and Death of the Solid South*, pp. 1–25; Woodward, *The Strange Career of Jim Crow*, pp. 60–109; Lawrence Goodwyn, *The Populist Moment: A Short History of the Agrarian Revolt in America* (New York: Oxford University Press, 1978); and Ayers, *The Promise of the New South*, pp. 214–82.

16. On disfranchisement, see Kousser, *The Shaping of Southern Politics*, pp. 224–65; Grantham, *The Life and Death of the Solid South*, pp. 1–25; and Walter Dean Burnham, "The System of 1896: An Analysis," in Paul Kleppner et al., eds., *The Evolution of American Electoral Systems* (Westport, CT: Greenwood Press, 1982), pp. 147–202.

17. See Grantham, *The Life and Death of the Solid South*, pp. 26–57. On the "white primary," see V. O. Key, Jr., *Southern Politics in State and Nation* (New York: Knopf, 1950), pp. 619–43.

18. See Key, *Southern Politics*, pp. 489–663.

19. Ibid., pp. 3–12.

20. Ibid., pp. 5–6.

21. Ibid., p. 307.

22. Ibid., pp. 298–311.

23. See ibid., pp. 254–76. See also Robert A. Caro, *The Years of Lyndon Johnson: The Path to Power* (London: Collins, 1983), pp. 3–49, 306–40.

24. On the Texas populist tradition, see Chandler Davidson, *Race and Class in Texas Politics* (Princeton, NJ: Princeton University Press, 1990), pp. 17–39.

25. On Long's Louisiana, see T. Harry Williams, *Huey Long* (New York: Vintage Books, 1981); and the more critical assessment in William Ivy Hair, *The Kingfish and His Realm: The Life and Times of Huey P. Long* (Baton Rouge: Louisiana State University Press, 1991). On Earl Long and the tradition of "Longism," see A. J. Liebling, *The Earl of Louisiana* (Baton Rouge: Louisiana State University Press, 1970); and Michael L. Kurtz and Morgan D. Peoples, *Earl K. Long: The Saga of Uncle Earl and Louisiana Politics* (Baton Rouge: Louisiana State University Press, 1990).

26. On Democratic dominance in presidential elections in the South, see Key, *Southern Politics*, pp. 315–44, 385–405; Grantham, *The Life and Death of the Solid South*, pp. 58–77; and Everett Carll Ladd, Jr., with Charles D. Hadley, *Transformations of the Party System: Political Coalitions from the New Deal to the 1970s*, 2nd ed. (New York: Norton, 1978), pp. 42–46.

27. On the history of the unit rule, see Howard L. Reiter, *Selecting the President: The Nominating Process in Transition* (Philadelphia: University of Pennsylvania Press, 1985), pp. 133–36.

28. On the two-thirds rule see ibid., pp. 133–34; and Earl Black and Merle Black, *The Vital South: How Presidents Are Elected* (Cambridge, MA: Harvard University Press, 1992), pp. 79–99. On the disastrous 1924 convention and the divisions in the Democratic party during the 1920s, see Robert K. Murray, *The 103rd Ballot: Democrats and the Disaster at Madison Square Garden* (New York: Harper & Row, 1976), esp. pp. 3–34.

29. In 1924, 55 percent of House Democrats and 54 percent of Senate Democrats came from the South. See Norman J. Ornstein, Thomas E. Mann, and Michael J. Malbin, *Vital Statistics on Congress: 1989–90* (Washington, DC: American Enterprise Institute, 1990), p. 11, p. 15.

30. On the voting records of southern Democrats in Congress during the 1920s, see Barbara Sinclair, *Congressional Realignment: 1925–1978* (Austin: University of Texas Press, 1982), pp. 18–50; and Richard Franklin Bensel, *Sectionalism and American Political Development: 1880–1980* (Madison: University of Wisconsin Press, 1984), pp. 128–47.

31. On the 1910 revolt and the committee seniority system, see James L. Sundquist, *The Decline and Resurgence of Congress* (Washington, DC: Brookings Institution, 1981), pp. 155–95; Ronald M. Peters, Jr., *The American Speakership: The Office in Historical Perspective* (Baltimore: Johns Hopkins University Press, 1990), pp. 75–145; and Barbara Sinclair, *The Transformation of the U.S. Senate* (Baltimore: Johns Hopkins University Press, 1989), pp. 8–29.

32. William S. White, *Citadel: The Story of the U.S. Senate* (New York: Harper Bros., 1956), p. 68.

33. On the southern cohesion in both houses of Congress, see Key, *Southern Politics*, pp. 345–82; and W. Wayne Shannon, "Revolt in Washington: The South

in Congress," in William C. Havard, ed., *The Changing Politics of the South* (Baton Rouge: Louisiana State University Press, 1972), pp. 637–87.

34. See Kousser, *The Shaping of Southern Politics*, pp. 11–44; and Grantham, *The Life and Death of the Solid South*, pp. 8–9.

35. On southern Republicans during the Solid South era, see Key, *Southern Politics*, pp. 277–97.

36. On the 1928 "bolt," see Key, *Southern Politics*, pp. 317–29.

37. On the South and the New Deal, see Nancy J. Weiss, *Farewell to the Party of Lincoln: Black Politics in the Age of FDR* (Princeton, NJ: Princeton University Press, 1983), pp. 157–79; Ladd and Hadley, *Transformations of the American Party System*, pp. 42–46; Grantham, *The Life and Death of the Solid South*, pp. 102–24; Bensel, *Sectionalism and American Political Development*, pp. 147–74.

38. Weiss, *Farewell to the Party of Lincoln*, pp. 96–135, 240–49.

39. On the founding of the conservative coalition, see James T. Patterson, *Congressional Conservatism and the New Deal: The Growth of the Conservative Coalition in Congress* (Lexington: University of Kentucky Press, 1967), pp. 77–187.

40. Bensel, *Sectionalism and American Political Development*, p. 176.

41. On black realignment during the 1930s, see Weiss, *Farewell to the Party of Lincoln*, pp. 180–235.

42. See John Morton Blum, *V Was for Victory: Politics and American Culture During World War II* (New York: Harvest Books, 1977), pp. 182–220.

43. On the "Dixiecrat" revolt and the 1948 election, see Key, *Southern Politics*, pp. 329–44.

44. On the emergence of the southern business class and the southern Republicans, see Earl Black and Merle Black, *Politics and Society in the South* (Cambridge, MA: Harvard University Press, 1987), pp. 23–72, 259–75; Jack Bass and Walter De Vries, *The Transformation of Southern Politics: Social Change and Political Consequence Since 1945* (New York: New American Library, 1977), pp. 3–40; Alexander P. Lamis, *The Two-Party South*, 2nd expanded ed. (New York: Oxford University Press, 1990), pp. 20–43; and Scher, *Politics in the New South*, pp. 96–191.

45. On migration, see Black and Black, *Politics and Society in the South*, pp. 3–22; and Paul Allen Beck, "Partisan Dealignment in the Post-War South," *American Political Science Review* 71 (1977): 477–96.

46. Lamis, *The Two-Party South*, pp. 20–43; Black and Black, *Politics and Society in the South*, pp. 259–75; and Bass and De Vries, *The Transformation of Southern Politics*, pp. 23–40.

47. Black and Black, *Politics and Society in the South*, pp. 276–91.

48. On the civil rights movement, see ibid., pp. 75–125; and Scher, *Politics in the New South*, pp. 195–326.

49. On Kennedy and civil rights, see Carl M. Brauer, *John F. Kennedy and the Second Reconstruction* (New York: Columbia University Press, 1977).

50. Ibid., pp. 230–310.

Chapter 3

1. On the pattern of the Wallace vote in 1968, see Kevin P. Phillips, *The Emerging Republican Majority* (Garden City, NY: Anchor Books, 1970), pp. 187–289; and Alexander P. Lamis, *The Two-Party South*, 2nd expanded ed. (New York: Oxford University Press, 1990), p. 29.

2. Phillips, *The Emerging Republican Majority*, pp. 248–49; and Lamis, *The Two-Party South*, p. 29.

3. Phillips, *The Emerging Republican Majority*, pp. 282–89.

4. Lamis, *The Two-Party South*, p. 29.

5. Earl Black and Merle Black, *Politics and Society in the South* (Cambridge, MA: Harvard University Press, 1987), pp. 237–45, 264–75; Lamis, *The Two-Party South*, pp. 20–43; Harold W. Stanley, "Southern Partisan Changes: Dealignment, Realignment or Both?" *Journal of Politics* 50 (1988): 64–88; and Norman H. Nie, Sidney Verba, and John R. Petrocik, *The Changing American Voter*, enlarged ed. (Cambridge, MA: Harvard University Press, 1979), pp. 217–23.

6. Black and Black, *Politics and Society in the South*, pp. 269–71; and Nie et al., *The Changing American Voter*, pp. 226–29.

7. On the 1976 election results, see Jack Bass and Walter De Vries, *The Transformation of Southern Politics: Social Change and Political Consequence Since 1945* (New York: New American Library, 1977), pp. 409–12.

8. Earl Black and Merle Black, *The Vital South: How Presidents Are Elected* (Cambridge, MA: Harvard University Press, 1992), p. 56.

9. Lamis, *The Two-Party South*, pp. 31–42; and Black and Black, *Politics and Society in the South*, p. 11.

10. Lamis, *The Two-Party South*, pp. 31–39.

11. Ibid., pp. 35–36.

12. Ibid., pp. 210–336.

13. Paul Allen Beck, "Partisan Dealignment in the Post-War South," *American Political Science Review* 71 (1977): 477–96; and Stanley, "Southern Partisan Changes," pp. 64–88.

14. On "split-level realignment," see Byron E. Shafer, "The Notion of an Electoral Order," in Byron E. Shafer, ed., *The End of Realignment? Interpreting American Electoral Eras* (Madison: University of Wisconsin Press, 1991), pp. 37–84; Everett Carll Ladd, Jr., "The 1988 Elections: Continuation of the Post-New Deal System," *Political Science Quarterly* 104 (1989): 1–18; and Michael Nelson, "Constitutional Aspects of the Elections," in Michael Nelson, ed., *The*

Elections of 1988 (Washington, DC: Congressional Quarterly Press, 1989), pp. 181–209.

15. On the abolition of the two-thirds rule, see Howard L. Reiter, *Selecting the President: The Nominating Process in Transition* (Philadelphia: University of Pennsylvania Press, 1985), p. 135; Nancy J. Weiss, *Farewell to the Party of Lincoln: Black Politics in the Age of FDR* (Princeton, NJ: Princeton University Press, 1983), p. 184; and Black and Black, *The Vital South*, pp. 90–92.

16. On the Mississippi Freedom Democratic party (MFDP), see Bass and De Vries, *The Transformation of Southern Politics*, pp. 203–7.

17. See Byron E. Shafer, *Quiet Revolution: The Struggle for the Democratic Party and the Shaping of Post-Reform Politics* (New York: Russell Sage Foundation, 1983), pp. 13–100, 523–39.

18. On the McGovern–Fraser reforms, see Shafer, *Quiet Revolution*; Nelson W. Polsby, *Consequences of Party Reform* (New York: Oxford University Press, 1983); and Austin Ranney, "The Political Parties: Reform and Decline" in Anthony King, ed., *The New American Political System* (Washington, DC: American Enterprise Institute, 1978), pp. 214–47.

19. Polsby, *Consequences of Party Reform*, pp. 53–64.

20. Ibid., pp. 72–78; and Reiter, *Selecting the President*, pp. 73–77. See also Byron E. Shafer, *Bifurcated Politics* (Cambridge, MA: Harvard University Press, 1978).

21. See Black and Black, *The Vital South*, pp. 241–71.

22. William G. Mayer, "The New Hampshire Primary: A Historical Overview," in Gary R. Orren and Nelson W. Polsby, eds., *Media and Momentum: The New Hampshire Primary and Nomination Politics* (Chatham, NJ: Chatham House, 1987), pp. 9–41.

23. On "momentum" and the New Hampshire primary, see William C. Adams, "As New Hampshire Goes . . . " in Orren and Polsby, eds., *Media and Momentum*, pp. 42–59.

24. On the strategy of the Carter campaign, see Jules Witcover, *Marathon: The Pursuit of the Presidency 1972–1976* (New York: Viking Press, 1977), pp. 105–38.

25. On Carter and blacks, see ibid., p. 337; and Lamis, *The Two-Party South*, pp. 37–39.

26. Bass and De Vries, *The Transformation of Southern Politics*, pp. 409–12; and Lamis, *The Two-Party South*, pp. 37–39. See also Black and Black, *The Vital South*, pp. 327–43.

27. On Wallace and Carter in 1976, see Witcover, *Marathon*, pp. 253–73.

28. On the surge in Republican support among white southerners, and particularly evangelical Christians, in the late 1970s, see Kevin P. Phillips, *Post-Conservative America: People, Politics & Ideology in a Time of Crisis* (New York: Vintage Books, 1983), pp. 90–91, 188–92.

29. On the collapse of Carter's white southern support in 1980, see William Schneider, "The November 4 Vote for President: What Did It Mean?" in Austin Ranney, ed., *The American Elections of 1980* (Washington, DC: American Enterprise Institute, 1981), pp. 212–62; and Black and Black, *The Vital South*, pp. 307–15.

30. On the 1984 primaries in the South, see Gerald Pomper, "The Nominations," in Gerald Pomper, ed., *The Election of 1984: Reports and Interpretations* (Chatham, NJ: Chatham House, 1985), pp. 1–34 esp. pp. 11–23; Nelson Polsby, "The Democratic Nomination and the Evolution of the Party System," in Austin Ranney, ed., *The American Elections of 1984* (Washington, DC: American Enterprise Institute, 1985), pp. 36–65; and Black and Black, *The Vital South*, pp. 256–60.

31. James R. Dickenson, "New Democratic Chairman Meets with the Party's Southern Leaders," *Washington Post*, 17 February 1985, p. A9.

32. On the formation of the DLC, see Jack W. Germond and Jules Witcover, *Whose Broad Stripes and Bright Stars? The Trivial Pursuit of the Presidency 1988* (New York: Warner Books, 1989), pp. 39–41; and Dan Balz and David S. Border, "The Rift in the Democratic Party Grows Wider," *Washington Post National Weekly Edition*, 11 March 1985, p. 13.

33. Michael Barone, "How Democrats, the Free-Trade Party, Got to Be Protectionists," *Washington Post National Weekly Edition*, 14 October 1985, pp. 23–24.

34. On the 1986 Senate elections in the South, see Germond and Witcover, *Whose Bright Stripes and Broad Stars?* pp. 45–48.

35. On the genesis of Super Tuesday 1988, see ibid., pp. 41–42; Harold W. Stanley and Charles D. Hadley, "The Southern Presidential Primary: Regional Intentions with National Implications," *Publius* 17 (1987): 83–100; Ronald Brownstein, "Moving Up," *National Journal*, 1 March 1986, p. 530.

36. Several observers noted the flaws in the Super Tuesday strategy: Jack Germond and Jules Witcover, "South's Regional Primary May Not Be Decisive," *National Journal*, 8 March 1986, p. 590; Alan Ehrenhalt, "Democrats Wooing Dixie Face Catch-22," *Congressional Quarterly Weekly Report*, 8 February 1986, p. 291; William Schneider, "South's Primaries: A Hollow Shell?" *National Journal*, 22 February 1986, pp. 470–71; and David S. Broder, "No More Super Tuesdays," *Washington Post*, 2 March 1988, p. A21.

37. On the effects of New Hampshire momentum on the Super Tuesday results, see Charles D. Hadley and Harold W. Stanley, "Super Tuesday 1988: Regional Results and National Implications," *Publius* 19 (1989): 19–37.

38. On Gore's strategy, see Thomas B. Edsall, "South's Leaders, Not Followers, Embracing Gore," *Washington Post*, 5 February 1988, p. A14.

39. On the Gephardt campaign, see Germond and Witcover, *Whose Bright Stripes and Broad Stars?* pp. 244–66, 282–86.

40. James R. Dickenson, "Nunn Decides Not to Seek White House," *International Herald Tribune*, 28 August 1987, p. 2.

41. For data on the 1988 Super Tuesday Voting, see Rhodes Cook, "One Side Is Clearer, the Other Murky," *Congressional Quarterly Weekly Report*, 12 March 1988, pp. 636–46.

42. On the absence of conservative Democrats from the polls on Super Tuesday, see Black and Black, *The Vital South*, pp. 260–71; Hadley and Stanley, "Super Tuesday 1988," pp. 19–37; and Craig Allen Smith and Kathy B. Smith, "Myths About Presidential Campaigning in the South: Dramatic Myths vs. Empirical Hypotheses in 1988," paper presented at the Citadel Symposium on Southern Politics, Charleston, SC, 9 March 1990.

43. Cook, "One Side Is Clearer."

44. Ibid. On the false expectations created by Dukakis's southern showing on Super Tuesday, see Germond and Witcover, *Whose Bright Stripes and Broad Stars?* pp. 289–90.

45. On Gore's disastrous New York and Wisconsin campaigns, see ibid., pp. 311–18.

Chapter 4

1. See James L. Sundquist, *The Decline and Resurgence of Congress* (Washington, DC: Brookings Institution, 1981), pp. 367–414; Samuel C. Patterson, "The Semi-Sovereign Congress," in Anthony King, ed., *The New American Political System* (Washington, DC: American Enterprise Institute, 1978), pp. 125–77; Lawrence C. Dodd and Bruce I. Oppenheimer, "Consolidating Power in the House: The Rise of a New Oligarchy," in Lawrence C. Dodd and Bruce I. Oppenheimer, eds., *Congress Reconsidered*, 4th ed. (Washington, DC: Congressional Quarterly Press, 1989), pp. 39–64; and Ronald M. Peters, Jr., *The American Speakership: The Office in Historical Perspective* (Baltimore: Johns Hopkins University Press, 1990), pp. 209–86.

2. V. O. Key, Jr., *Southern Politics in State and Nation* (New York: Knopf, 1949), pp. 369–82.

3. Ibid.; J. Morgan Kousser, *The Shaping of Southern Politics: Suffrage Restriction and the Establishment of the One-Party South 1880–1910* (New Haven, CT: Yale University Press, 1974); and C. Vann Woodward, *The Strange Career of Jim Crow*, 3rd ed. (New York: Oxford University Press, 1974).

4. See Key, *Southern Politics*, pp. 345–82; and Barbara Sinclair, *Congressional Realignment, 1925–1978* (Austin: University of Texas Press, 1978), pp. 18–72.

5. See James T. Patterson, *Congressional Conservatism and the New Deal: The Growth of the Conservative Coalition in Congress* (Lexington: University of Kentucky Press, 1967).

6. See James MacGregor Burns, *The Deadlock of Democracy: Four Party Politics in America* (Englewood Cliffs, NJ: Prentice-Hall, 1963), pp. 301–22; W. Wayne Shannon, "Revolt in Washington: The South in Congress," in William C. Havard, ed., *The Changing Politics of the South* (Baton Rouge: Louisiana State University Press, 1972), pp. 637–87; and Richard Franklin Bensel, *Sectionalism and American Political Development: 1880–1980* (Madison: University of Wisconsin Press, 1984), pp. 147–255.

7. On the southern electoral realignment, see Jack Bass and Walter De Vries, *The Transformation of Southern Politics: Social Change and Political Consequence Since 1945* (New York: Meridian Books, 1977), pp. 3–55; Alexander P. Lamis, *The Two-Party South*, 2nd expanded ed. (New York: Oxford University Press, 1990), pp. 3–43; Earl Black and Merle Black, *Politics and Society in the South* (Cambridge, MA: Harvard University Press, 1987), pp. 232–16. For an economic interpretation of the southern realignment, see Bensel, *Sectionalism and American Political Development*, pp. 175–255.

8. On the 1970s reforms and their effects, see Sundquist, *The Decline and Resurgence of Congress*, pp. 367–414.

9. Interview with Congressman Jones, 11 July 1990.

10. Interview with Congressman Hutto, 10 July 1990.

11. Interview with Congressman Darden, 21 June 1990.

12. Interview with Congressman Derrick, 26 June 1990.

13. Interview with Congressman Clement, 18 July 1990.

14. Interview with Congressman Bennett, 18 June 1990.

15. Interview with Congressman Pickett, 19 June 1990.

16. Interview with Congressman Browder, 20 June 1990.

17. Interview with Congressman Anthony, 19 July 1990.

18. Interview with Congressman Harris, 24 July 1990.

19. Interview with Congresswoman Patterson, 27 June 1990.

20. Interview with Congressman Spratt, 22 June 1990.

21. Interview with Congressman Neal, 10 July 1990.

22. Interview with Congressman Payne, 27 June 1990.

23. Interview with Congressman Lewis, 20 June 1990.

24. Interview with Congressman Espy, 20 June 1990.

25. Interview with Congressman Jenkins, 20 July 1990.

26. Parker interview.

27. Payne interview.

28. Darden interview.

29. Interview with Congressman Huckaby, 20 July 1990.

30. Interview with Congressman Barnard, 19 June 1990.

31. Interview with Congressman Stenholm, 22 June 1990.

32. Interview with Congressman Leath, 26 July 1990.

33. Jenkins interview.

34. Interview with Congressman Ireland, 26 July 1990.

35. See Key, *Southern Politics*, p. 367; and Bensel, *Sectionalism and American Political Development*, pp. 147–74.

36. See David W. Rhode, " 'Something's Happening Here; What It Is Ain't Exactly Clear': Southern Democrats in the House of Representatives," in Morris P. Fiorina and David W. Rhode, eds., *Home Style and Washington Work: Studies of Congressional Politics* (Ann Arbor: University of Michigan Press, 1989), pp. 137–63.

37. See Lamis, *The Two-Party South*, pp. 210–36.

38. Lewis interview.

39. Interview with Congressman Tallon, 27 June 1990.

40. Payne interview.

41. Interview with Congressman Hatcher, 26 June 1990.

42. Darden interview.

43. Interview with Congressman Hayes, 12 July 1990.

44. Jenkins interview.

45. Parker interview.

46. Spratt interview.

47. Barnard interview.

48. Neal interview.

49. Payne interview.

50. Hayes interview.

51. Hutto interview.

52. Browder interview.

53. Clement interview.

54. Barnard interview.

55. Pickett interview.

56. Espy interview.

57. Interview with Congressman Price, 24 July 1990.

58. See Rhode, " 'Something's Happening Here,' " pp. 137–63.

59. See *National Journal*, 17 June 1989, pp. 1445–46. After Congressman Gray announced his retirement in June 1991, another race for the position of House Democratic whip ensued between Congressman David Bonior of Michigan and Congressman Steny Hoyer from the border state of Maryland. Bonior won by 160 to 109 votes.

60. Jenkins interview.

61. Anthony interview.

62. Hayes interview.

63. Interview, anonymity requested.

64. Interview, anonymity requested.

65. Interview, anonymity requested.

66. Huckaby interview.

67. Neal interview.

68. Spratt interview.
69. Hutto interview.
70. Interview, anonymity requested.
71. Interview, anonymity requested.
72. Interview, anonymity requested.
73. Jones interview.
74. Spratt interview.
75. Stenholm interview; and Rhode, " 'Something's Happening Here,' " pp. 137–63.
76. Anthony interview.
77. Jenkins interview.
78. Bennett interview.
79. Hatcher interview.
80. Parker interview.
81. Neal interview.
82. Patterson interview.
83. Hatcher interview.
84. Payne interview.
85. On the boll weevils and the 1981 Reagan budget, see *National Journal*, 28 February 1981, pp. 350–54; and *Congressional Quarterly Weekly Report*, 18 April 1981, p. 672; 13 June 1981, pp. 1023–26; 27 June 1981, pp. 1127–28; 1 August 1981, pp. 1371–74; and 19 December 1981, p. 2555.
86. Stenholm interview.
87. Leath interview.
88. *Congressional Quarterly Weekly Report*, 1 August 1981, p. 1128.
89. Leath interview.
90. Stenholm interview.
91. Stenholm interview.
92. Hatcher interview.
93. Tallon interview.
94. Hayes interview.
95. Jenkins interview.
96. Hayes interview.
97. Huckaby interview.
98. Interview with Congressman Valentine, 27 June 1990.
99. Leath interview.
100. Hutto interview.
101. Harris interview.
102. Only three of the thirty districts held by the southern Democratic members interviewed were *not* carried by the Republican presidential candidate George Bush in 1988: Georgia's Fifth District (John Lewis); Mississippi's Second District (Mike Espy); and North Carolina's Second District (Tim Valentine).

103. Browder interview.

104. Espy interview. In the 1990 elections, Congressman Espy won with 84 percent of the vote!

105. Interview, anonymity requested.

106. Parker interview.

107. Bennett interview.

108. Barnard interview.

109. Hayes interview.

110. Interview, anonymity requested.

111. See William S. White, *Citadel: The Story of the U.S. Senate* (New York: Harper Bros., 1957), pp. 67–79; and Donald R. Matthews, *U.S. Senators and Their World* (New York: Vintage Books, 1960), pp. 92–117.

112. On the southerners' use of Senate rules and procedures to defeat or weaken civil rights legislation in the 1940s, see Gilbert C. Fite, *Richard B. Russell, Jr., Senator from Georgia* (Chapel Hill: University of North Carolina Press, 1991), pp. 224–42.

113. On the formation of the conservative coalition, see Patterson, *Congressional Conservatism and the New Deal*, pp. 77–163.

114. Ibid., p. 132. On the essential compromise that kept the Democratic coalition together in Congress during the New Deal era, see also Bensel, *Sectionalism and American Political Development*, pp. 157–255.

115. Interview with former Senator Long, 18 July 1990.

116. Interview with former Senator Fulbright, 19 July 1990.

117. On the struggle of the 1957 and 1960 Civil Rights acts, see Fite, *Richard B. Russell, Jr.*, pp. 329–48. On southern resistance to the *Brown* decision, see Dewey W. Grantham, *The Life and Death of the Solid South: A Political History* (Lexington: University Press of Kentucky, 1988), pp. 125–48.

118. On the impact of the "class of 1958" and the transformation of the Senate, see Barbara Sinclair, *The Transformation of the U.S. Senate* (Baltimore: Johns Hopkins University Press, 1989), pp. 30–50; and Michael Foley, *The New Senate: Liberal Influence on a Conservative Institution 1959–72* (New Haven, CT: Yale University Press, 1980).

119. Long interview.

120. See Foley, *The New Senate*, pp. 170–230.

121. On the New South Democrats, see Lamis, *The Two-Party South*, pp. 20–43.

122. On increased partisanship in the Senate, see Barbara Sinclair, "The Congressional Party: Evolving Organizational, Agenda-Setting and Policy Roles," in L. Sandy Maisel, ed., *The Parties Respond: Changes in the American Party System* (Boulder, CO: Westview Press, 1990), pp. 227–48.

123. Interview with Senator Breaux, 25 July 1990.

124. Interview with Senator Shelby, 28 June 1990.

125. Interview with Garrett, press secretary to Senator Johnston, 16 July 1990.

126. Shelby interview.

127. Breaux interview.

128. Interview with Bull, legislative director to Senator Bentsen, 19 July 1990.

129. On southern Democratic senators and the Bork nomination, see Ethan Bronner, *Battle for Justice: How the Bork Nomination Shook America* (New York: Norton, 1989), pp. 277–306.

130. Garrett interview.

131. Breaux interview.

132. Shelby interview.

133. Interview with Fleming, administrative assistant to Senator Ford, 29 June 1990.

134. Bull interview.

135. Garrett interview.

136. Interview with Poole, state director for Senator Sanford, 11 July 1990.

137. Breaux interview.

138. Fleming interview.

139. Garret Interview.

140. Breaux interview.

141. Garrett interview.

Chapter 5

1. Anthony Downs, *An Economic Theory of Democracy* (New York: Harper & Row, 1957).

2. See James Q. Wilson, *The Amateur Democrat: Club Politics in Three Cities* (Chicago: University of Chicago Press, 1962).

3. See Nelson W. Polsby, *Consequences of Party Reform* (New York: Oxford University Press, 1983); and Austin Ranney, "The Political Parties: Reform and Decline," in Anthony King, ed., *The New American Political System* (Washington, DC: American Enterprise Institute, 1978), pp. 213–47.

4. See Nicol C. Rae, *The Decline and Fall of the Liberal Republicans: From 1952 to the Present* (New York: Oxford University Press, 1989).

5. On the Ripon Society and the liberal Republican failure, see ibid., pp. 78–121.

6. Interview with Al From, 18 July 1990. "Superdelegate" refers to the delegate slots at the national convention (15 percent of the total), which were set aside for Democratic party officeholders and elected officials (including members of Congress, governors, state chairs, and mayors) by the Hunt Commission (on the party's presidential nominating process) in 1983. On the founding of the DLC, see also Jon F. Hale, "The Institutionalization of a Party Faction: The Case

of the Democratic Leadership Council," paper presented at the annual meeting of the Midwest Political Science Association, Chicago, 18–20 April 1991.

7. Interview with Senator Robb, 28 June 1990; From interview; and Dom Bonafede, "Kirk at the DNC's Helm," *National Journal*, 22 March 1986, pp. 703–7.

8. Richard E. Cohen, "Democratic Leadership Council Sees Party Void and Is Ready to Fill It," *National Journal*, 1 February 1986, pp. 267–70.

9. Ibid.

10. From interview.

11. On the concept of liberal fundamentalism, see William Galston and Elaine Ciulla Kamarck, *The Politics of Evasion: Democrats and the Presidency* (Washington, DC: Progressive Policy Institute, 1989). For a critique, see Robert Kuttner, "What's the Beef? The Once and Future DLC," *New Republic*, 2 April 1990, pp. 16–19.

12. Robb interview.

13. Bonafede, "Kirk at the DNC's Helm."

14. Ibid.

15. From interview.

16. For an excellent analysis of the reasons for the Democrats' loss in 1988, see Galston and Kamarck, *The Politics of Evasion*. In regard to the Democrats' problem with the white suburban middle class, see Peter Brown, *Minority Party: Why Democrats Face Defeat in 1992 and Beyond* (Washington, DC: Regnery-Gateway, 1991).

17. Interview with Bruce Reed, policy director of the DLC, 2 July 1991.

18. *New Democrat*, May 1991.

19. Reed interview; David S. Broder, "Democratic Leaders Complete the Groundwork for Change," *Miami Herald*, 12 May 1991, p. 3C; and David S. Broder, "The White Men in Suits Move In," *Economist*, 11 May 1991, pp. 21–22.

20. Reed interview.

21. *The New Orleans Declaration: A Democratic Agenda for the 1990s* (Washington, DC: Democratic Leadership Council, 1990).

22. Ibid.

23. On the Cleveland conference, see Broder, "Democratic Leaders"; Broder, "The White Men in Suits"; Gwen Ifill, "Democratic Group in Dispute over Goals," *New York Times*, 7 May 1991, p. A21; and Robin Toner, "Centrist Democrats Set Agenda for '92," *New York Times*,, 8 May 1991, p. A10.

24. Broder, "Democratic Leaders"; Broder, "The White Men in Suits."

25. Broder, "The White Men in Suits."

26. On the Ripon Society's impact, see Rae, *The Decline and Fall of the Liberal Republicans*, pp. 81–86, 99–102.

27. Galston and Kamarck, *The Politics of Evasion*.

28. From interview.

29. Ibid.

30. DLC membership list.

31. Broder, "Democratic Leaders." See also Ifill, "Democratic Group"; and Broder, "The White Men in Suits."

32. On the Coalition for Democratic Values, see Thomas B. Edsall, "Democrats, Seeking Shelter from Desert Storm," *Washington Post National Weekly Edition*, 4–10 February 1991, p. 14; and Robin Toner, "Liberals in Search of Values Run into Discord over War," *New York Times*, 27 January 1991, p. 11.

33. Robin Toner, "Adding up Concern for Bush and Doubts About Quayle, the Democrats Get Zero," *New York Times*, 12 May 1991, p. 1E.

34. Robert Kuttner, *The Life of the Party: Democratic Prospects in 1988 and Beyond* (New York: Penguin Books, 1988); and Kevin Phillips, *The Politics of Rich and Poor: Wealth and the American Electorate in the Reagan Aftermath* (New York: Random House, 1990).

35. Kuttner, "What's the Beef?" p 19.

36. For an excellent critique of the Phillips book, see Thomas Byrne Edsall, "The Hidden Role of Race," *New Republic*, 30 July and 6 August 1990, pp. 35–40.

37. From interview. The reference is to Michael Barone, *Our Country: The Shaping of America from Roosevelt to Reagan* (New York: Free Press, 1990), pp. xi–xvi.

38. Daniel Bell, *The End of Ideology: On the Exhaustion of Political Ideas in the Fifties* (Cambridge, MA: Harvard University Press, 1988), pp. 393–447.

39. Seymour Martin Lipset, *Political Renewal on the Left: A Comparative Perspective* (Washington, DC: Progressive Policy Institute, 1990), p. 21.

40. For a good account of how campaign-financing reforms and dependence on business political action committees have weakened the Democratic party's economic populism, see Thomas Byrne Edsall, *The New Politics of Inequality* (New York: Norton, 1984).

41. Ibid., pp. 25–26.

42. For an excellent case study of how the liberals lost the white working class, see Jonathan Rieder, *Canarsie: The Jews and Italians of Brooklyn Against Liberalism* (Cambridge, MA: Harvard University Press, 1985). On the same theme, see also Brown, *Minority Party*; Thomas Byrne Edsall and Mary D. Edsall, *Chain Reaction: The Impact of Race, Rights and Taxes on American Politics* (New York: Norton, 1991); James Davison Hunter, *Culture Wars: The Struggle to Define America* (New York: Basic Books, 1991); and Steven M. Gillon, *The Democrats' Dilemma: Walter F. Mondale and the Liberal Legacy* (New York: Columbia University Press, 1992).

43. For an account of how the Republicans' general antipathy for government damages them in congressional elections, see Alan Ehrenhalt, *The United States of Ambition* (New York: Random House, 1991), pp. 208–27.

44. See E. J. Dionne, Jr., *Why American Hate Politics* (New York: Simon & Schuster, 1991).

Chapter 6

1. The phenomenon of "declinism" that ultimately filtered down from elite levels to mass consciousness was initially inspired by Paul Kennedy's very influential best-seller, *The Rise and Fall of the Great Powers: Economic Change and Military Conflict from 1500 to 2000* (New York: Random House, 1989).

2. Leading Washington commentator David S. Broder captured the prevailing wisdom regarding Democratic prospects for the presidency in the immediate aftermath of the Gulf War, in an article entitled "The White House Is Almost a Lock, So What About Congress? Bush Could Help the GOP Win It All in '92," *Washington Post National Weekly Edition*, 15–21 April 1991, p. 21.

3. On Jackson's motivations regarding the 1992 race, see Steven A. Holmes, "Jackson May Not Be Running but He's Still a Front-Runner," *New York Times*, 27 June 1991, p. A1; and Robin Toner, "In Spotlight, Jackson Nears Decision," *New York Times*, 15 September 1991, p. 18.

4. On Harkin, see Robin Toner, "Senator Harkin Joins Race for President," *New York Times*, 16 September 1991, p. A10; and Sidney Blumenthal, "The Primal Scream, Tom Harkin and the Democratic Id," *New Republic*, 21 October 1991, pp. 22–25. Harkin's candidacy had the effect of nullifying the impact of the usually significant Iowa caucuses; see R. W. Apple, Jr., "No Suspense as Caucuses Near in Iowa," *New York Times*, 13 January 1992, p. A13.

5. On Brown, see Gwen Ifill, "Gov. Brown Makes Third Run for President," *New York Times*, 22 October 1992, p. A7; and B. Drummond Ayres, Jr., "Brown Hopes One-Note Campaign Strikes a Responsive Chord," *New York Times*, 28 December 1991, p. 7.

6. On Bradley's problems in New Jersey, see Michael Barone and Grant Ujifusa, *The Almanac of American Politics 1992* (Washington, DC: National Journal, Inc., 1991), pp. 772–74.

7. See Robin Toner, "Rockefeller's Assets," *New York Times Magazine*, 21 July 1992, p. 19.

8. On the early positive coverage of Kerrey, see Robin Toner, "The Unfinished Politician," *New York Times Magazine*, 14 April 1991, p. 43.

9. On the later negative Kerrey coverage, see Robert Pear, "Kerrey, Health-Care Plan Author Has Few of His Workers Covered," *New York Times*, 28 December 1991, p. 1; Elizabeth Kolbert, "Both Kerreys Searching for Campaign Strategy," *New York Times*, 16 January 1992, p. A10; and Sidney Blumenthal, "The Politics of Self," *New Republic*, 20 January 1992, pp. 22–27.

10. Steven Pearlstein, "Piecing Together the Puzzle of Paul Tsongas," *Washington Post National Weekly Edition*, 29 April–5 May 1991, pp. 8–9; and Karen

De Witt, "Tsongas Pitches Economic Austerity, Mixed with Patriotism," *New York Times*, 1 January 1992, p. 8.

11. On Cuomo's procrastination over entering the race, see Maureen Dowd, "To Run or Not to Run? For Cuomo, The Quandary Continues to Govern," *New York Times*, 4 November 1991, p. B1.

12. Kevin Sack, "Cuomo Says He Will Not Run for President in '92," *New York Times*, 21 December 1991, p. 1.

13. On Nunn's problems, see "The Politics of War," *Southern Political Report*, 22 January 1991; "The Spoils of Victory," *Southern Political Report*, 5 March 1991; and Robin Toner, "For First Time, Nunn's in Fray for Opposing One," *New York Times*, 13 March 1991, p. A12.

14. On Robb's problems, see B. Drummond Ayres, Jr., "Eavesdropping Tiff Entrances Virginia," *New York Times*, 9 June 1991, p. 12; and Kent Jenkins, Jr., "A Chink in the Armor of the Upright Charles Robb," *Washington Post National Weekly Edition*, 30 September–6 October, 1991, p. 13.

15. Gwen Ifill, "Gephardt Makes It Official: He's No '92 Candidate," *New York Times*, 18 July 1990, p. A14.

16. On Wilder, see Steven A. Holmes, "Wilder Seeks to Mix Black Support with Middle-Class Vote," *New York Times*, 30 December 1991, p. A10; and B. Drummond Ayres, Jr., "Wilder Ends Race for Presidency, Citing Virginia's Fiscal Troubles," *New York Times*, 9 January 1992, p. A1.

17. On Gore, see Robin Toner, "The Field Is Still Open, Time Wanes and, as in '88, the Name Gore Arises," *New York Times*, 25 March 1991, p. A9; Dan Balz, "How to Get out Front Without Really Running," *Washington Post National Weekly Edition*, 29 April–5 May 1991, pp. 14–15; and Gwen Ifill, "Gore Won't Run for President in 1992," *New York Times*, 22 August 1991, p. A16.

18. On Clinton's record in Arkansas, see Peter Applebome, "Clinton Record in Leading Arkansas: Successes, but Not Without Criticism," *New York Times*, 22 December 1991, p. 30.

19. Dan Balz, "Bill Clinton Has a New Script for the Democratic Repertory," *Washington Post National Weekly Edition*, 15–21 July 1991, p. 13; Anthony Lewis, "Arkansas Traveler," *New York Times*, 25 September 1991, p. A13; and Morton Kondracke, "Slick Willy: Bill Clinton, Postliberal Man," *New Republic*, 21 October 1991, pp. 18–21.

20. See Robert S. Boyd, "Wilder's Exit Helps Clinton Maneuver to Front of Pack," *Miami Herald*, 10 January 1992, p 1A.

21. See Gwen Ifill, "Trying to Deliver a Mainstream Message with a Southern Accent," *New York Times*, 27 December 1992, p. A12.

22. See Sidney Blumenthal, "The Anointed: Bill Clinton, Nominee-Elect," *New Republic*, 3 February 1992, pp. 24–27.

23. On Clinton's "southern strategy," see Gwen Ifill, "Clinton Moves on 2 Fronts to Widen Support in South," *New York Times*, 22 January 1992, p. A12.

The "southern primary" in 1992 was staggered, with Georgia voting on 3 March, South Carolina on 7 March, and Florida, Louisiana, Mississippi, Tennessee, and Texas on March 10. The remaining southern states voted later in the season. This arrangement diluted the influence of New Hampshire and enhanced that of Georgia, thus assisting the prospects of any southern regional contender.

24. On Carville, Begala, and the Wofford race, see Michael deCourcy Hinds, "Wofford Win Shows Voter Mood Swing," *New York Times*, 7 November 1991, p. A11.

25. On Clinton's front-runner status, see Dan Balz and E. J. Dionne, Jr., "Gathering Momentum, Money and Media Scrutiny Clinton Breaks out of the Democratic Pack," *Washington Post National Weekly Edition*, 20–26 January 1992, p. 12.

26. See Robin Toner, "Tsongas Surges as Voters Focus on Economics," *New York Times*, 10 February 1992, p. A10; and Maureen Dowd, "Surging in New Hampshire, Tsongas Is the First to Marvel," *New York Times*, 16 February 1992, p. 1.

27. See Richard L. Berke, "Off Course, Harkin Searches for the Main Campaign Highway," *New York Times*, 1 February 1992, p. 8. Allessandra Stanley, "Kerrey on the Run, Pursued by Failure," *New York Times*, 25 February 1992, p. A16; and Sidney Blumenthal, "Brownian Motion: The Other Jerry's Telethon," *New Republic*, 2 March 1992, pp. 18–20.

28. On Tsongas's liabilities, see Sidney Blumenthal, "The Puritan: Paul Tsongas's Will to Power," *New Republic*, 23 March 1992, pp. 10–12.

29. On Clinton's triumph in the Florida straw poll, see Tom Fiedler, "Florida Delivers for Clinton," *Miami Herald*, 16 December 1991, p. 1A.

30. See Gwen Ifill, "Clinton Defends His Privacy and Says the Press Intruded," *New York Times*, 27 January 1992, p. A8; and Gwen Ifill, "Clinton Attempts to Ignore Rumors," *New York Times*, 28 January 1992, p. A10.

31. See Gwen Ifill, "Clinton Thanked Colonel in '69 for 'Saving Me from the Draft'," *New York Times*, 13 February 1992, p. A1; and David E. Rosenbaum, "Clinton Could Have Known Draft Was Unlikely for Him," *New York Times*, 14 February 1992, p. A1.

32. See Robin Toner, "Democratic Race in New Hampshire Becoming Fiercer," *New York Times*, 12 February 1992, p. A1.

33. See Gwen Ifill, "Clinton Finds a Theme: He Is Not Paul Tsongas," *New York Times*, 5 March 1992, p. A11.

34. See R. W. Apple, Jr., "Washington Suburbs Offer Tsongas a Chance to Win Another," *New York Times*, 27 February 1992, p. A13.

35. See Ronald Smothers, "Blacks Feeling Like Wall Flowers as No Candidates Come Courting," *New York Times*, 2 March 1992, p. A1.

36. See Jeffrey Schmalz, "In Florida, Democrats Face Critical Electorate," *New York Times*, 9 March 1992, p. A10.

37. See Robert Reinhold, "No Longer Target of Jokes, Brown Becomes a Force," *New York Times*, 9 March 1992, p. A1.

38. See Richard L. Berke, "Tsongas Plays Role of David Opposing a Goliath, Clinton," *New York Times*, 10 March 1992, p. A16; and *Southern Political Report*, 19 February 1992.

39. On Tsongas's problems in the rust belt, see Richard L. Berke, "Tsongas Tells Auto Workers of Problem with Democrats," *New York Times*, 13 March 1992, p. A11; and Sidney Blumenthal, "The Pol: Bill Clinton in Illinois," *New Republic*, 6 April 1992, pp. 16–19.

40. On Brown's pitch to organized labor, see Jeffrey Schmalz, "Licking Bruises, Car Workers Are Hearing Brown Message," *New York Times*, 13 March 1992, p. A1. On Clinton's strategy, see Blumenthal, "The Pol"; Robin Toner, "Illinois: After Laying the Groundwork in 1991, Clinton Hopes for a Decisive Victory," *New York Times*, 13 March 1992, p. A11; and R. W. Apple, Jr., "Clinton Building Strong Coalition, Leads in Michigan," *New York Times*, p. A1.

41. Robin Toner, "Tsongas Abandons Campaign, Leaving Clinton a Clear Path Toward Showdown with Bush," *New York Times*, 20 March 1992, p. A1.

42. See Robin Toner, "Clinton's Foes Ask If He's Electable," *New York Times*, 17 March 1992, p. A1.

43. See Lally Weymouth, "The Once and Future Presidential Candidate: Dick Gephardt Waits in the Wings," *Washington Post National Weekly Edition*, 9–15 March 1992, p. 25; and Stewart L. Udall, "Who's Presidential? The Strongest Democrats Aren't Running," *New York Times*, 12 March 1992, p. A23.

44. See R. W. Apple, Jr., "The Longest of Long Shots Is Suddenly a Contender," *New York Times*, 29 March 1992, p. E1. On Brown as a vehicle for the anti-Clinton forces, see Howell Raines, "Nail-biting Time Nears for Democratic Party," *New York Times*, 31 March 1992, p. A11; and Elizabeth Kolbert, "Brown Is Helped by Clinton Foes," *New York Times*, 2 April, 1992, p. A10.

45. See Robin Toner, "New York Battle Gains in Intensity as Brown Surges," *New York Times*, 26 March 1992, p. A1; and Elizabeth Kolbert, "Clinton Strategy Changes, Putting Brown on Defense," *New York Times*, 4 April 1992, p. 1.

46. See Maureen Dowd, "Candidate Is Tripped up over Alliance with Jackson," *New York Times*, 3 April 1992, p. A8.

47. See Isabel Wilkerson, "In Land of Underdogs, Brown Advantage Slips," *New York Times*, 6 April 1992, p. A10.

48. On Tsongas's flirtation with reentry, see R. W. Apple, Jr., "Campaign Fury Is Giving Way to Speculation," *New York Times*, 7 April 1992, p. A13.

49. On Perot's impact on the Republicans in his spring campaign, see Thomas C. Hayes, "In Sharpest Attack, Perot Accuses Bush of Inaction," *New York Times*, 24 April 1992, p. A10; Robin Toner, "Anxious Days for Bush's Campaign as GOP Heads into a 3-Way Race," *New York Times*, 21 May 1992, p. A10; and

Fred Barnes, "The Spoiler: Will Perot Hurt or Help Clinton?" *New Republic*, 15 June 1992, pp. 10–11.

50. On the platform, see Robert Pear, "In a Final Draft, Democrats Reject a Part of Their Past," *New York Times*, 26 June 1992, p. A9.

51. On Clinton and Jackson, see Sidney Blumenthal, "The Reanointed," *New Republic*, 27 July 1992, pp. 10–14; and David S. Broder, "Jackson's Decline, Brown's Ascent," *Washington Post National Weekly Edition*, 20–26 July 1992, p. 4. On the marginalization of the New Left faction of the party at the convention, see Thomas B. Edsall, "The Democrats Pick a New Centerpiece," *Washington Post National Weekly Edition*, 20–26 July 1992, p. 14.

52. On the Gore selection and his impact on the ticket, see Dan Balz, "The Party Is Betting Youth Is the Ticket," *Washington Post National Weekly Edition*, 13–19 July 1992, p. 8.

53. On Clinton's postconvention surge in the polls against Bush, see Howard Kurtz, "On a Roll with a Bounce Riding a Wave: Clinton and Gore Are Soaring on the Media Roller Coaster," *Washington Post National Weekly Edition*, 3–9 August 1992, p. 16.

54. See Sidney Blumenthal, "Party Time: The Rise of the Democrats," *New Republic*, 10 August 1993, pp. 16–17.

55. On Clinton's emphasis on the economy in the general election, see "The War Room Drill," *Newsweek* (special election issue), November–December 1992, pp. 78–81.

56. On the GOP convention, see Thomas B. Edsall, "The Republicans' Value-added Strategy: Zeroing in on Cultural, Sexual and Racial Themes," *Washington Post National Weekly Edition*, 24–30 August 1992, p. 15; David S. Broder, "On a Right Wing and a Prayer," *Washington Post National Weekly Edition*, 31 August–6 September 1992, p. 4; E. J. Dionne, Jr., "Having Some Second Thoughts: The GOP Is Reassessing Its Family Values Emphasis," *Washington Post National Weekly Edition*, 31 August–6 September 1992, p. 15; and Andrew Rosenthal, "Bush Tries to Recoup from Harsh Tone on 'Values'," *New York Times*, 21 September 1992, p. A1.

57. On the erosion of the Republican electoral coalition, see Robin Toner, "Clinton Retains Significant Lead in Latest Survey: Democrat Runs Strong Among Major Parts of Reagan's Bloc," *New York Times*, 16 September 1992, p. A1; and Thomas B. Edsall, "Bloc Busting: The Demise of the GOP Quest for a Majority," *Washington Post National Weekly Edition*, 19–25 October 1992, p. 2. On the failure of foreign policy to work for Bush in 1992, see Fred Barnes, "Churchill Syndrome," *New Republic*, 26 October 1992, pp. 12–13.

58. See Robin Toner, "Perot Re-Enters the Campaign, Saying Bush and Clinton Fail to Address Government 'Mess'," *New York Times*, 2 October 1992, p. A1.

59. See Andrew Rosenthal, "Bush Campaign Welcomes Perot as Reconfiguring the Election," *New York Times*, 2 October 1992, p. A10.

60. "Face to Face in Prime Time," *Newsweek* (special election issue), November–December 1992, pp. 88–91; and Sidney Blumenthal, "Why Am I Here?" *New Republic*, 9 November 1992, pp. 16–18.

61. See Robin Toner, "Polls Say Clinton Keeps Lead Despite Furious GOP Fire," *New York Times*, 1 November 1992, p. A1.

62. On the relationship between the rise in turnout and the Perot candidacy, see Robert Pear, "55% Voting Rate Reverses 30-Year Decline," *New York Times*, 5 November 1992, p. B4; and Adam Clymer, "Turnout on Election Day '92 Was the Largest in 24 Years," *New York Times*, 17 December 1992, p. A13.

63. The information on the voting behavior comes from the results of the Voter Research and Surveys (VRS) poll, conducted by the major news organizations. For the results, see "Portrait of the Electorate," *New York Times*, 5 November 1992, p. B9.

64. On the second choices of Perot voters, see Steven A. Holmes, "An Eccentric but No Joke: Perot''s Strong Showing Raises Questions on What Might Have Been, and Might Be," *New York Times*, 5 November 1992, p. A1.

65. See Jeffrey Schmalz, "Gay Areas Are Jubilant over Clinton," *New Your Times*, 5 November 1992, p. B16.

66. On how the "gays in the military" issue reopened the Democratic party's ideological fault lines, see Michael Kelley, "President's Early Troubles Rooted in Party's Old Strains," *New York Times*, 2 February 1992, p. A1.

67. Zoe Baird ultimately withdrew her nomination for attorney general after the revelation that she and her husband had hired illegal aliens to take care of their child. After the withdrawal of Clinton's second choice, Judge Kimba Wood of New York, because of a similar problem, the post was eventually filled by Janet Reno, state attorney of Dade County, Florida.

68. See Catherine S. Manegold, "Clinton's Ire Pushes Women to Anger and Reflection," *New York Times*, 23 December 1992, p. A11; and Gwen Ifill, "As Pressure Grows, Clinton Tries to Fill out Cabinet," *New York Times*, 24 December 1992, p. A9.

69. See Fred Barnes, "Neoconned: Clinton and the Hawks," *New Republic*, 25 January 1992, pp. 14–16.

BIBLIOGRAPHY

Interviews

Congressman Beryl F. Anthony, Jr., 19 July 1990.
Congressman Doug Barnard, Jr., 19 June 1990.
Congressman Charles E. Bennett, 18 June 1990.
Senator John B. Breaux, 25 July 1990.
Congressman Glen Browder, 20 June 1990.
Blaine Bull, legislative director to Senator Lloyd Bentsen, 19 July 1990.
Congressman Bob Clement, 18 July 1990.
Congressman George "Buddy" Darden, 21 June 1990.
Congressman Butler C. Derrick, 26 June 1990.
Congressman Glenn English, 28 June 1990.
Congressman Mike Espy, 20 June 1990.
James Fleming, administrative assistant to Senator Wendell Ford, 29 June 1990.
Al From, executive director, Democratic Leadership Council, 18 July 1990.
Senator J. William Fulbright, 19 July 1990.
Tony Garrett, press secretary to Senator Bennett Johnston, 16 July 1990.
Congressman Sam M. Gibbons, 25 July 1990.
Congressman Claude Harris, Jr., 24 July 1990.
Congressman Charles F. Hatcher, 26 June 1990.
Congressman James A. Hayes, 12 July 1990.
Congressman Jerry Huckaby, 20 July 1990.
Congressman Earl Hutto, 10 July 1990.
Congressman Andy Ireland, 26 July 1990.
Congressman Ed Jenkins, 20 July 1990.
Congressman Walter B. Jones, 11 July 1990.
Gene Karp, administrative assistant to Senator Dennis DeConcini, 15 June 1990.
Congressman J. Marvin Leath, 26 July 1990.
Congressman John R. Lewis, 20 June 1990.
Senator Russell B. Long, 18 July 1990.
Congressman Stephen L. Neal, 10 July 1992.

Norman J. Ornstein, resident scholar at the American Enterprise Institute, 25 June 1990.
Congressman Mike Parker, 11 July 1990.
Congresswoman Elizabeth J. Patterson, 27 June 1990.
Congressman Lewis F. Payne, 27 June 1990.
Congressman Timothy J. Penny, 17 June 1990.
Congressman Owen B. Pickett, 19 June 1990.
Sam Poole, state director to Senator Terry Sanford, 11 July 1990.
Congressman David E. Price, 24 July 1990.
Bruce Reed, policy director, Democratic Leadership Council, 2 July 1990.
Senator Charles S. Robb, 28 June 1990.
Peter Rosenblatt, president, Coalition for a Democratic Majority, 2 July 1990 (by telephone).
Senator Richard C. Shelby, 28 June 1990.
Congressman Ike Skelton, 26 June 1990.
Congressman John M. Spratt, 22 June 1990.
Congressman Richard H. Stallings, 16 July 1990.
Congressman Charles W. Stenholm, 22 June 1990.
Congressman Robin M. Tallon, 27 June 1990.
Congressman Tim Valentine, 27 June 1990.

Newspapers and Periodicals

Congressional Quarterly Weekly Report
Economist
International Herald Tribune
Miami Herald
National Journal
New Democrat
New Republic
New York Review of Books
New York Times
Southern Political Report
Washington Post

Books and Articles

Adams, William C. "As New Hampshire Goes. . . ." In Nelson W. Polsby and Gary R. Orren, eds., *Media and Momentum: The New Hampshire Primary and Nomination Politics*, pp. 42–59. Chatham, NJ: Chatham House, 1987.

Agee, James, and Walker Evans. *Let Us Now Praise Famous Men: Three Tenant Families*. Boston: Houghton Mifflin, 1988.

Allswang, John M. *Bosses, Machines and Urban Voters*. Rev. ed. Baltimore: Johns Hopkins University Press, 1986.

Ayers, Edward L. *The Promise of the New South: Life After Reconstruction*. New York: Oxford University Press, 1992.

Baltzell, E. Digby. *The Protestant Establishment: Aristocracy and Caste in America*. New Haven, CT: Yale University Press, 1964.

Barone, Michael. *Our Country: The Shaping of America from Roosevelt to Reagan*. New York: Free Press, 1990.

Barry, John M. *The Ambition and the Power*. New York: Viking Press, 1989.

Bass, Jack, and Walter De Vries. "Cross Pressures in the White South." In Seymour Martin Lipset, ed., *Emerging Coalitions in American Politics*, pp. 307–24. San Francisco: Institute for Contemporary Studies, 1978.

——. *The Transformation of Southern Politics: Social Change and Political Consequence Since 1945*. New York: New American Library, 1977.

Beck, Paul Allen. "Partisan Dealignment in the Post-War South." *American Political Science Review* 71 (1977): 477–96.

Bell, Daniel. *The End of Ideology: On the Exhaustion of Political Ideas in the Fifties*. Cambridge, MA: Harvard University Press, 1988.

Beller, Dennis C., and Frank P. Belloni, eds. *Faction Politics: Political Parties and Factionalism in Comparative Perspective*. Santa Barbara, CA: ABC–Clio, 1978.

Bensel, Richard Franklin. *Sectionalism and American Political Development: 1880–1980*. Madison: University of Wisconsin Press, 1984.

Bernick, E. Lee, Patricia K. Freeman, and David M. Olson. "Southern State Legislatures: Recruitment and Reform." In James F. Lea, ed., *Contemporary Southern Politics*, pp. 214–41. Baton Rouge: Louisiana State University Press, 1988.

Billington, Monroe Lee. *The Political South in the 20th Century*. New York: Scribner, 1975.

Black, Earl and Merle Black. *Politics and Society in the South. Cambridge, MA: Harvard University Press, 1987.*

——. *The Vital South: How Presidents Are Elected*. Cambridge, MA: Harvard University Press, 1992.

Boorstin, Daniel J. *The Americans: The Colonial Experience*. New York: Vintage Books, 1958.

——. *The Americans: The Democratic Experience*. New York: Vintage Books, 1974.

——. *The Americans: The National Experience*. New York: Vintage Books, 1965.

188   *Bibliography*

Brady, David W. "Coalitions in the U.S. Congress." In L. Sandy Maisel, ed., *The Parties Respond: Changes in the American Party System*, pp. 249–66. Boulder, CO: Westview Press, 1990.

———. *Critical Elections and Congressional Policy Making*. Stanford, CA: Stanford University Press, 1988.

Brogan, Hugh. *The Pelican History of the United States of America*. Harmondsworth: Penguin Books, 1986.

Bronner, Ethan. *Battle for Justice: How the Bork Nomination Shook America*. New York, Norton, 1989.

Brown, Peter. *Minority Party: Why Democrats Face Defeat in 1992 and Beyond*. Washington, DC: Regnery-Gateway, 1991.

Bullock, Charles S., III. "The Nomination Process and Super Tuesday. In Laurance W. Moreland, Robert P. Steed, and Tod A. Baker, eds., *The 1988 Presidential Election in the South: Continuity Amidst Change*, pp. 3–19. New York: Praeger, 1991.

———. "The South in Congress: Power and Policy." In James F. Lea, ed., *Contemporary Southern Politics*, pp. 177–93. Baton Rouge: Louisiana State University Press, 1988.

Burnham, Walter Dean. *Critical Elections and the Mainsprings of American Politics*. New York: Norton, 1970.

———. "Critical Realignment: Dead or Alive." In Byron E. Shafer, ed., *The End of Realignment: Interpreting America's Electoral Eras*, pp. 101–39. Madison: University of Wisconsin Press, 1991.

———. *The Current Crisis in American Politics*. New York: Oxford University Press, 1982.

———. "The System of 1896: An Analysis." In Paul Kleppner et al., eds., *The Evolution of American Electoral Systems*, pp. 147–202. Westport, CT: Greenwood Press, 1982.

Carmines, Edward G., and James A. Stimson. *Issue Evolution: Race and the Transformation of American Politics*. Princeton, NJ: Princeton University Press, 1989.

Caro, Robert A. *The Years of Lyndon Johnson: Means of Ascent*. New York: Knopf, 1990.

———. *The Years of Lyndon Johnson: The Path to Power*. London: Collins, 1983.

Cash, W.J. *The Mind of the South*. New York: Vintage Books, 1969.

Cook, Rhodes. "The Nominating Process." In Michael Nelson, ed., *The Elections of 1988*, pp. 25–61. Washington, DC: Congressional Quarterly Press, 1989.

Dauer, Manning J. "Florida: The Different State." In William C. Havard, ed., *The Changing Politics of the South*, pp. 92–164. Baton Rouge: Louisiana State University Press, 1972.

Davidson, Chandler. *Race and Class in Texas Politics*. Princeton, NJ: Princeton University Press, 1990.

Dionne, E.J., Jr. *Why Americans Hate Politics*. New York: Simon & Schuster, 1991.

Downs, Anthony. *An Economic Theory of Democracy*. New York: Harper & Row, 1957.

Edsall, Preston W., and J. Oliver Williams. "North Carolina: Bipartisan Paradox." In William C. Havard, ed., *The Changing Politics of the South*, pp. 366–423. Baton Rouge: Louisiana State University Press, 1972.

Edsall, Thomas Byrne. *The New Politics of Inequality*. New York: Norton, 1984.

Edsall, Thomas Byrne, and Mary D. Edsall. *Chain Reaction: The Impact of Race, Rights, and Taxes on American Politics*. New York: Norton, 1991.

Ehrenhalt, Alan. *The United States of Ambition*. New York: Random House, 1991.

Eisenberg, Ralph N. "Virginia: The Emergence of Two-Party Politics." In William C. Havard, ed., *The Changing Politics of the South*, pp. 39–91. Baton Rouge: Louisiana State University Press, 1972.

Eubanks, Cecil L. "Contemporary Southern Politics: Present State and Future Possibilities." In James F. Lea, ed., *Contemporary Southern Politics*, pp. 287–99. Baton Rouge: Louisiana State University Press, 1988.

Fischer, David Hackett. *Albion's Seed: Four British Folkways in America*. New York: Oxford University Press, 1989.

Fite, Gilbert C. *Richard B. Russell, Jr., Senator from Georgia*. Chapel Hill: University of North Carolina Press, 1991.

Foley, Michael. *The New Senate: Liberal Influence on a Conservative Institution*. New Haven, CT: Yale University Press, 1980.

Frady, Marshall. *Wallace*. Revised ed. New York: New American Library, 1976.

Furgurson, Ernest B. *Hard Right: The Rise of Jesse Helms*. New York: Norton, 1986.

Galston, William, and Elaine Ciulla Kamarck. *The Politics of Evasion: Democrats and the Presidency*. Washington, DC: Progressive Policy Institute, 1989.

Germond, Jack W., and Jules Witcover. *Whose Broad Stripes and Bright Stars? The Trivial Pursuit of the Presidency, 1988*. New York: Warner Books, 1989.

Gillon, Steven M. *The Democrats' Dilemma: Walter F. Mondale and the Liberal Legacy*. New York: Columbia University Press, 1992.

Goodwyn, Lawrence. *The Populist Moment: A Short History of the Agrarian Revolt in America*. New York: Oxford University Press, 1978.

Grantham, Dewey W. *The Life and Death of the Solid South: A Political History*. Lexington: University Press of Kentucky, 1988.

Gunther, John. *Inside USA*. London: Hamish Hamilton, 1947.

Hadley, Charles W. and Harold W. Stanley. "The Southern Presidential Primary: Regional Results and National Implications." *Publius* 19 (1989): 19–37.

Hagstrom, Jerry. *Beyond Reagan: The New Landscape of American Politics*. New York: Norton, 1988.

Hale, Jon F. "The Institutionalization of a Party Faction: The Case of the Democratic Leadership Council." Paper presented at the annual meeting of the Midwest Political Science Association, Chicago, 18–20 April 1991.

Hartz, Louis. *The Liberal Tradition in America*. New York: Harcourt Brace Jovanovich, 1955.

Havard, William C., ed. *The Changing Politics of the South*. Baton Rouge: Louisiana State University Press, 1972.

Hofstadter, Richard. *The American Political Tradition and the Men Who Made It*. New York: Vintage Books, 1973.

Huckshorn, Robert J., ed. *Government and Politics in Florida*. Gainsville: University of Florida Press, 1991.

Hunter, James Davison. *Culture Wars: The Struggle to Define America*. New York: Basic Books, 1991.

Jacobson, Gary C. *The Electoral Origins of Divided Government: Competition in U.S. House Elections, 1946–1988*. Boulder, CO: Westview Press, 1990.

Kamarck, Elaine Ciulla. "Structure as Strategy: Presidential Nominating Politics in the Post-Reform Era." In L. Sandy Maisel, ed., *The Parties Respond: Changes in the American Party System*, pp. 160–86. Boulder, CO: Westview Press, 1990.

Key, V.O., Jr. *Southern Politics: In State and Nation*. New York: Knopf, 1949.

Kleppner, Paul. "From Ethno-Religious Conflict to 'Social Harmony': Coalitional and Party Transformations in the 1890s." In Seymour Martin Lipset, ed., *Emerging Coalitions in American Politics*, pp. 41–59. San Francisco: Institute for Contemporary Studies, 1978.

Kleppner, Paul, et al., eds. *The Evolution of American Electoral Systems*. Westport, CT: Greenwood Press, 1982.

Kousser, J. Morgan. *The Shaping of Southern Politics: Suffrage Restriction and the Establishment of the One-Party South, 1880–1910*. New Haven, CT: Yale University Press, 1974.

Kuttner, Robert. *The Life of the Party: Democratic Prospects in 1988 and Beyond*. New York: Penguin Books, 1988.

Ladd, Everett Carll, Jr., "Like Waiting for Godot: The Uselessness of Realignment for Understanding Change in Contemporary American Politics." In Byron E. Shafer, ed., *The End of Realignment: Interpreting America's Electoral Eras*, pp. 24–36. Madison: University of Wisconsin Press, 1991.

——. "The 1988 Elections: Continuation of the Post-New Deal System." *Political Science Quarterly* 104 (1989): 1–18.

———. "The Shifting Party Coalitions: 1932–1976." In Seymour Martin Lipset, ed., *Emerging Coalitions in American Politics*, pp. 81–102. San Francisco: Institute for Contemporary Studies, 1978.

———. With Charles D. Hadley. *Transformations of the American Party System: Political Coalitions from the New Deal to the 1970s*. 2nd ed. New York: Norton, 1978.

Lamis, Alexander P. *The Two-Party South*. 2nd expanded ed. New York: Oxford University Press, 1990.

Lasch, Christopher. *The True and Only Heaven: Progress and Its Critics*. New York: Norton, 1991.

Lea, James F., ed. *Contemporary Southern Politics*. Baton Rouge: Louisiana State University Press, 1988.

Leuchtenberg, William E. *Franklin D. Roosevelt and the New Deal: 1932–1940*. New York: Harper & Row, 1963.

Leyburn, James G. *The Scotch-Irish: A Social History*. Chapel Hill: University of North Carolina Press, 1962.

Liebling, A.J. *The Earl of Lousiana*. Baton Rouge: Louisiana State University Press, 1970.

Link, Arthur S. *Woodrow Wilson and the Progressive Era: 1910–1917*. New York: Harper Torchbooks, 1963.

Lipset, Seymour Martin, ed. *Emerging Coalitions in American Politics*. San Francisco, Institute for Contemporary Studies, 1978.

Lubell, Samuel. *The Future of American Politics*. 3rd ed. New York: Harper & Row, 1965.

Luebke, Paul. *Tar Heel Politics: Myths and Realities*. Chapel Hill: University of North Carolina Press, 1990.

Maginnis, John. *Cross to Bear*. Baton Rouge: Darkhorse Press, 1992.

———. *The Last Hayride*. Baton Rouge, LA: Gris Gris Press, 1984.

Maisel, L. Sandy, ed. *The Parties Respond: Changes in the American Party System*. Boulder, CO: Westview Press, 1990.

Matthews, Donald R. *U.S. Senators and Their World*. New York: Vintage Books, 1960.

Mayer, William G. "The New Hampshire Primary: A Historical Overview." In Nelson W. Polsby and Gary R. Orren, eds., *Media and Momentum: The New Hampshire Primary and Nomination Politics*, pp. 9–37. Chatham, NJ: Chatham House, 1987.

Mayhew, David R. *Placing Parties in American Politics*. Princeton, NJ: Princeton University Press, 1986.

McGerr, Michael E. *The Decline of Popular Politics: The American North, 1986–1928*. New York: Oxford University Press, 1986.

McKeever, Porter. *Adlai Stevenson: His Life and Legacy*. New York: Morrow, 1989.

McPherson, James M. *Battle Cry of Freedom: The Civil War Era*. New York: Ballantine Books, 1988.

Moreland, Laurence W., Robert P. Steed, and Tod A. Baker, eds. *The 1988 Presidential Election in the South: Continuity Amidst Change in Southern Politics*. New York: Praeger, 1991.

Munger, Frank, and James Blackhurst. "Factionalism in the National Conventions, 1940–1964: An Analysis of Ideological Consistency in State Delegation Voting." *Journal of Politics* 27 (1965): 375–94.

Murray, Robert K. *The 103rd Ballot: Democrats and the Disaster at Madison Square Garden*. New York: Harper & Row, 1976.

Naipaul, V.S. *A Turn in the South*. New York: Knopf, 1989.

Nelson, Michael, ed. *The Elections of 1988*. Washington, DC: Congressional Quarterly Press, 1989.

Nie, Norman H., Sidney Verba, and John R. Petrocik. *The Changing American Voter*. Enlarged ed. Cambridge, MA: Harvard University Press, 1979.

Ornstein, Norman J., Thomas E. Mann, and Michael J. Malbin. *Vital Statistics on Congress: 1989–90*. Washington, DC: American Enterprise Institute, 1990.

Orren, Gary R. "Candidate Style and Voter Alignment in 1976." In Seymour Martin Lipset, ed., *Emerging Coalitions in American Politics*, pp. 127–81. San Francisco: Institute for Contemporary Studies, 1978.

Orren, Gary R., and Nelson W. Polsby, eds. *Media and Momentum: The New Hampshire Primary and Nomination Politics*. Chatham, NJ: Chatham House, 1987.

Patterson, James T. *Congressional Conservatism and the New Deal: The Growth of the Conservative Coalition in Congress*. Lexington: University of Kentucky Press, 1967.

Percy, William Alexander. *Lanterns on the Levee: Recollections of a Planter's Son*. New York: Knopf, 1941.

Peters, Ronald M., Jr. *The American Speakership: The Office in Historical Perspective*. Baltimore: Johns Hopkins University Press, 1990.

Phillips, Kevin P. *The Emerging Republican Majority*. Garden City, NY: Anchor Books, 1970.

———. *The Politics of Rich and Poor: Wealth and the American Electorate in the Reagan Aftermath*. New York: Random House, 1990.

———. *Post-Conservative America: People, Politics and Ideology in a Time of Crisis*. New York: Vintage Books, 1983.

Pomper, Gerald. "The Nominations." In Gerald Pomper, ed., *The Election of 1984: Reports and Interpretations*, pp. 1–34. Chatham, NJ: Chatham House, 1985.

Rae, Nicol C. *The Decline and Fall of the Liberal Republicans: From 1952 to the Present*. New York: Oxford University Press, 1989.

Ranney, Austin. "The Political Parties: Reform and Decline." In Anthony King, ed., *The New American Political System*, pp. 213–47. Washington, DC: American Enterprise Institute, 1978.

Reed, John Shelton. *The Enduring South: Subcultural Persistence in a Mass Society*, 2nd ed. Chapel Hill: University of North Carolina Press, 1986.

Reichley, A. James. *Conservatives in an Age of Change: The Nixon and Ford Administrations*. Washington, DC: Brookings Institution, 1981.

——. *The Life of the Parties: A History of American Political Parties*. New York: Free Press, 1992.

——. *Religion in American Public Life*. Washington, DC: Brookings Institution, 1985.

Reiter, Howard L. "Intra-Party Cleavages in the United States Today." *Western Political Quarterly* 34 (1981): 287–300.

——. *Selecting the President: The Nominating Process in Transition*. Philadelphia: University of Pennsylvania Press, 1985.

Reston, James, Jr. *The Lone Star: The Life of John Connally*. New York: Harper & Row, 1989.

Rhode, David W. " 'Something's Happening Here; What It Is Ain't Exactly Clear': Southern Democrats in the House of Representatives." In Morris P. Fiorina and David W. Rhode, eds., *Home Style and Washington Work: Studies of Congressional Politics*, pp. 137–63. Ann Arbor: University of Michigan Press, 1989.

Rieder, Jonathan. *Canarsie: The Jews and Italians of Brooklyn Against Liberalism*. Cambridge, MA: Harvard University Press, 1985.

Roback, Thomas H., and Judson L. James. "Party Factions in the United States." In Dennis C. Beller and Frank P. Belloni, eds., *Faction Politics: Political Parties and Factionalism in Comparative Perspective*, pp. 329–55. Santa Barbara, CA: ABC–Clio, 1978.

Rose, Richard. *The Problem of Party Government*. London: Macmillan, 1974.

Rossiter, Clinton. *Conservatism in America*. London: Heinemann, 1955.

Sabato, Larry. "New South Governors and the Governorship." In James F. Lea, ed., *Contemporary Southern Politics*, pp. 194–213. Baton Rouge: Louisiana State University Press, 1988.

Sartori, Giovanni. *Parties and Party Systems: A Framework for Analysis*. Vol. 1. Cambridge: Cambridge University Press, 1976.

Scammon, Richard M., and Ben J. Wattenberg. *The Real Majority: An Extraordinary Examination of the American Electorate*. New York: Coward, McCann & Geoghegan, 1970.

Scher, Richard K. *Politics in the New South: Republicanism, Race and Leadership in the 20th Century*. New York: Paragon House, 1992.

Schlesinger, Arthur M., Jr., *The Age of Jackson*. London: Eyre & Spottiswoode, 1946.

194   *Bibliography*

Schneider, William. "Democrats and Republicans, Liberals and Conservatives."
    In Seymour Martin Lipset, ed., *Emerging Coalitions in American Poli-
    tics*, pp. 183–267. San Francisco: Institute for Contemporary Studies,
    1978.
Shafer, Byron E. *Bifurcated Politics: Evolution and Reform in the National Party
    Convention*. Cambridge, MA: Harvard University Press, 1988.
——. *Is America Different? A New Look at American Exceptionalism*. Oxford:
    Oxford University Press, 1991.
——. "The Notion of an Electoral Order: The Structure of Electoral Politics at
    the Accession of George Bush." In Byron E. Shafer, ed., *The End of
    Realignment: Interpreting America's Electoral Eras*, pp. 37–84. Madison:
    University of Wisconsin Press, 1991.
——. *Quiet Revolution: The Struggle for the Democratic Party and the Shaping of
    Post-Reform Politics*. New York: Russell Sage Foundation, 1983.
Shannon, W. Wayne. "Revolt in Washington: The South in Congress." In
    William C. Havard, ed., *The Changing Politics of the South*, pp. 637–87.
    Baton Rouge: Louisiana State University Press, 1972.
Silbey, Joel H. "Beyond Realignment and Realignment Theory: American Politi-
    cal Eras, 1789–1989." In Byron E. Shafer, ed., *The End of Realignment:
    Interpreting America's Electoral Eras*, pp. 3–23. Madison: University of
    Wisconsin Press, 1991.
——. *The Partisan Imperative: The dynamics of American Politics Before the
    Civil War*. New York: Oxford University Press, 1985.
——. "The Rise and Fall of American Political Parties." In L. Sandy Maisel, ed.,
    The Parties Respond: Changes in the American Party System, pp. 3–17.
    Boulder, CO: Westview Press, 1990.
Sinclair, Barbara. "The Congressional Party: Evolving Organizational, Agenda-
    Setting, and Policy Roles." In L. Sandy Maisel, ed., *The Parties Re-
    spond: Changes in the American Party System*, pp. 227–48. Boulder, CO:
    Westview Press, 1990.
——. *Congressional Realignment, 1925–1978*. Austin: University of Texas
    Press, 1982.
——. *The Transformation of the U.S. Senate*. Baltimore, Johns Hopkins Univer-
    sity Press, 1989.
Snider, William D. *Helms and Hunt: The North Carolina Senate Race, 1984*.
    Chapel Hill: University of North Carolina Press, 1985.
Stanley, Harold W. "Southern Partisan Changes: Dealignment, Realignment of
    Both?" *Journal of Politics* 50 (1988): 64–88.
Stanley, Harold W., and Charles W. Hadley. "The Southern Presidential Primary:
    Regional Results with National Implications." *Publius* 17 (1987): 83–100.
Sundquist, James L. *The Decline and Resurgence of Congress*. Washington, DC:
    Brookings Institution, 1981.

———. *Dynamics of the Party System: Alignment and Realignment of Political Parties in the United States*. Revised ed. Washington, DC: Brookings Institution, 1983.

Thompson, Michael, Richard Ellis, and Aaron Wildavsky. *Cultural Theory*. Boulder, CO: Westview Press, 1990.

"Twelve Southerners." *I'll Take My Stand: The South and the Agrarian Tradition*. Baton Rouge: Louisiana State University Press, 1977.

Warren, Robert Penn. *All The King's Men*. New York: Harcourt Brace Jovanovich, 1974.

Weaver, Richard M. *The Southern Tradition at Bay: A History of Postbellum Thought*. Washington, DC: Regnery-Gateway, 1989.

Weiss, Nancy J. *Farewell to the Party of Lincoln: Black Politics in the Age of FDR*. Princeton, NJ: Princeton University Press, 1983.

White, Theodore H. *The Making of the President 1960*. New York: Atheneum, 1961.

White, William S. *Citadel: The Story of the U.S. Senate*. New York: Harper Bros., 1956.

Wildavsky, Aaron. *The Revolt Against the Masses: And Other Essays on Politics and Public Policy*. New York: Basic Books, 1971.

Wilson, Charles Reagan, and William Ferris, eds. *Encyclopedia of Southern Culture*. Chapel Hill: University of North Carolina Press, 1989.

Wilson, James Q. *The Amateur Democrat: Club Politics in Three Cities*. Chicago: University of Chicago Press, 1962.

Witcover, Jules. *Marathon: The Pursuit of the Presidency, 1972–76*. New York: Viking Press, 1976.

Wood, Goron S. *The Creation of the American Republic, 1776–1787*. New York: Norton. 1972.

———. *The Radicalism of the American Revolution*. New York: Knopf, 1992.

Woodward, C. Vann. *The Strange Career of Jim Crow*. 3rd revised ed. New York: Oxford University Press, 1974.

Wyatt-Brown, Bertram. *Southern Honor: Ethics and Behaviour in the Old South*. Oxford: Oxford University Press, 1982.

INDEX